Building Code Basics: Fire

Based on the 2009 International Fire Code®

Building Code Basics: Fire

Based on the 2009 International Fire Code®

International Code Council
Scott Stookey

DELMAR
CENGAGE Learning™

Australia • Brazil • Japan • Korea • Mexico • Singapore • Spain • United Kingdom • United States

Building Code Basics: Fire, Based on the 2009 International Fire Code®
Scott Stookey

Vice President, Technology and Trades
Professional Business Unit:
Gregory L. Clayton

Product Development Manager:
Ed Francis

Development: Dawn M. Jacobson

Editorial Assistant: Nobina Chakraborti

Director of Marketing: Beth A. Lutz

Executive Marketing Manager:
Taryn Zlatin McKenzie

Marketing Manager: Marissa Maiella

Production Director: Carolyn Miller

Production Manager: Andrew Crouth

Senior Content Project Manager:
Kara A. DiCaterino

Art Director: Benjamin Gleeksman

ICC Staff:

Senior Vice President, Business and
Product Development: Mark A. Johnson

Technical Director, Product Development:
Doug Thornburg

Manager, Project and Special Sales:
Suzane Nunes Holten

Senior Marketing Specialist:
Dianna Hallmark

For product information and technology assistance, contact us at
**Professional Group Cengage Learning Customer &
Sales Support, 1-800-354-9706**

For permission to use material from this text or product,
submit all requests online at **cengage.com/permissions.**
Further permissions questions can be e-mailed to
permissionrequest@cengage.com.

Library of Congress Control Number: 2009932854
ISBN-13: 978-1-4354-0070-2
ISBN-10: 1-4354-0070-4

ICC World Headquarters
500 New Jersey Avenue, NW
6th Floor
Washington, D.C. 20001-2070
Telephone: 1-888-ICC-SAFE (422-7233)
Website: http://www.iccsafe.org

Delmar
5 Maxwell Drive
Clifton Park, NY 12065-2919
USA

Cengage Learning is a leading provider of customized learning solutions with office locations
around the globe, including Singapore, the United Kingdom, Australia, Mexico, Brazil and Japan.
Locate your local office at: **international.cengage.com/region**

Cengage Learning products are represented in Canada by Nelson Education, Ltd.

For more learning solutions, please visit our corporate website at
www. cengage.com.

Visit us at **www. InformationDestination.com**

NOTICE TO THE READER

Publisher does not warrant or guarantee any of the products described herein or perform any
independent analysis in connection with any of the product information contained herein.
Publisher does not assume, and expressly disclaims, any obligation to obtain and include information other than that provided to it by the manufacturer. The reader is expressly warned to
consider and adopt all safety precautions that might be indicated by the activities described
herein and to avoid all potential hazards. By following the instructions contained herein, the
reader willingly assumes all risks in connection with such instructions. The publisher makes no
representations or warranties of any kind, including but not limited to, the warranties of fitness
for particular purpose or merchantability, nor are any such representations implied with respect
to the material set forth herein, and the publisher takes no responsibility with respect to such
material. The publisher shall not be liable for any special, consequential, or exemplary damages
resulting, in whole or part, from the readers' use of, or reliance upon, this material.

Printed in the United States of America
1 2 3 4 5 XX 11 10 09

CONTENTS

PART 1: CODE ADMINISTRATION AND ENFORCEMENT

PART 2: GENERAL SAFETY REQUIREMENTS

PART 3: SITE AND BUILDING SERVICES

PART 4: FIRE/LIFE SAFETY SYSTEMS AND FEATURES

PART 5: SPECIAL PROCESSES AND BUILDING USES

PART 6: HAZARDOUS MATERIALS

PREFACE

Fire code enforcement is an important public safety function because unwanted fires kill and injure thousands annually. Unwanted fires have a monetary impact on communities because fires remove businesses from the tax rolls while the damaged building is rebuilt and reconstructed. Statistics confirm that over 40% of the businesses that experience a fire never reopen because they lose their customer base. Of concern to any community is the accidental release of hazardous materials because of their potential for fire, explosion, or causing injury due to incapacitation by the chemical's constituent. All of these incidents require a response by the fire department, which places fire fighters in danger, especially when an interior rescue and fire attack is required. Given the broad scope of hazards in society, the job of enforcing the fire code is challenging. This is especially true when dealing with hazardous materials, high-piled combustible storage, and combustible dust-producing operations.

Building Code Basics: Fire – Based on the 2009 International Fire Code® was developed to address the need for an illustrated text explaining the basics of the fire code. It is intended to provide an understanding of the proper application of the code to the most commonly encountered hazards found in many communities and cities. The text is presented and organized in a user friendly manner with an emphasis on technical accuracy and clear non-code language. The content is directed to fire service professionals, code officials, designers and others in the building construction industry.

The content of *Building Code Basics: Fire* is organized to correspond to the arrangement of the 2009 IFC. It commences with a review of the legal aspects associated with the adoption and enforcement of the fire code provisions including permitting, right of entry, and inspector liability. It progresses through common hazards that can be found in any occupancy, site and building features that must be addressed with any new construction, fire and life safety systems and features, special processes and uses, and it concludes with a review of the most commonly encountered hazardous materials. This format is useful to readers because it pulls together related information from the various sections of the IFC into one convenient location while providing a familiar frame of reference to those with any code enforcement experience. The arrangement of the book is formatted to follow the steps of new building construction or renovation as well as areas of focus during any fire inspection. This format and arrangement offers the reader a means to understanding why fire code enforcement is an important public safety function and why it is so important to the safety of emergency responders.

Anyone involved in the design, construction, or inspection of buildings or industrial processes and hazards will benefit from this book. Beginning and experienced fire inspectors, plans examiners, contractors, engineers, architects, environmental/health and safety professionals, and students in fire science, fire protection, and building inspection technology curriculum or related fields of study and work will gain a fundamental understanding and practical application of the frequently used provisions of the 2009 edition of the IFC.

Reasonable and correct application of the code provisions is enhanced by a basic understanding of the fire code development process, the scope, intent, and correlation of the family of the International Codes, and the proper administration of those codes. This fundamental information is provided in the opening chapters of this manuscript. The book also explains the interaction of the fire code with other local and state regulations. Because the content is focused on the fire code, the book includes prerequisite reading which is important in understanding the *International Building Code* occupancy classification system, how buildings are assigned occupancy classifications, and how these classifications are used in the application of the fire code.

This book does not intend to cover all provisions of the IFC or all of the accepted materials and methods for the construction of fire protection systems, features, or the storage and handling of combustible and hazardous materials. Focusing in some detail on the most common hazards that are found in nearly every community affords an opportunity to fully understand the basics without exploring every variable and alternative. This is not to say that information not covered is any less important or valid. This book is best used as a companion to the IFC and appropriate National Fire Protection Association standards, which should be referenced for more complete information.

Building Code Basics: Fire features full color illustrations and photographs to assist the reader in visualizing the application of the code requirements. Practical examples, simplified tables, and highlights of particularly useful information also aid in understanding the provisions and determining code compliance. References to the applicable 2009 IFC sections are cited to assist readers in locating the corresponding code language and related topics in the code.

ABOUT THE INTERNATIONAL FIRE CODE

The IFC is a comprehensive, stand-alone model code that regulates minimum fire safety requirements for new and existing buildings, facilities, storage, and processes. The IFC addresses fire prevention, fire protection, life safety, and safe storage and use of hazardous materials in new and existing buildings, facilities, and processes. The IFC provides a total approach of controlling hazards in all buildings and sites, regardless of the hazard being indoors or outdoors.

The IFC is a design document. For example, before one constructs a building, the site must be provided with an adequate water supply for fire-fighting operations and a means of building access for emergency responders in the event of a medical emergency, fire, or natural or technological disaster. Depending on the building's occupancy and uses, the IFC regulates the various hazards that may be housed within the building, including refrigeration systems, application of flammable finishes, fueling of motor vehicles, high-piled combustible storage, and the storage and use of hazardous materials. The IFC sets forth minimum requirements for these and other hazards and contains requirements for maintaining the life safety of building occupants, the protection of emergency responders, and to limit the damage to a building and its contents as the result of a fire, explosion, or unauthorized hazardous material discharge.

ACKNOWLEDGMENTS

Building Code Basics: Fire is the result of a collaborative effort, and the author is grateful for the valuable contributions by the following talented staff of ICC Business and Product Development: Hamid Naderi, P.E., C.B.O., Vice President, Doug Thornburg, A.I.A., C.B.O., Technical Director of Product Development, and Senior Technical Staff Stephen Van Note, C.B.O. and Peter Kulczyk, C.B.O. for their guidance and contributions throughout this project. Steve's experience helped make a very challenging project less difficult, and his guidance is appreciated. As always, Doug's guidance on various means of egress provisions, as well his experience, proved helpful.

The manuscript was peer-reviewed by Robert Neale, Director of National Fire Service Programs at the U.S. Fire Administration in Emmitsburg, MD. Robert is the consummate educator and fire code official and made significant contributions to this book with his numerous suggestions. He is a thorough reviewer and his technical analysis helped in making this into an exceptional book. The author extends his thanks to Robert for his timely, thorough review and constructive criticisms.

ABOUT THE AUTHOR

Scott Stookey is a Senior Technical Staff member with the International Code Council. Scott previously was the Fire Protection Engineer for the Phoenix (AZ) Fire Department's Special Hazard Unit and was a Engineering Associate with the Austin (TX) Fire Department Hazardous Materials Engineering section. He is a graduate of the Fire Protection and Safety Engineering Technology program at Oklahoma State University and has over 20 years of experience in the areas of regulatory compliance and emergency response.

PREREQUISITE READING— OCCUPANY CLASSIFICATION

Before readers of this book proceed into its content, they must understand that most communities regulate their buildings based on the occupancy classification. Occupancy is assigned based on the use and character of a building. A building's use is evaluated for life safety and fire risks, and its character represents the functions and activities that are expected to occur in the building. A correct occupancy classification establishes the foundation for all the code requirements that are intended for the building's safe use.

Occupancies are classified into groups and sub-groups using the requirements in the *International Building Code* (IBC). Normally the fire code official does not have the legal authority in most communities to assign an occupancy classification—this task is assigned to the building code official. The reason is the IBC has requirements that address not only fire and life safety aspects, but includes requirements for accessibility of mobility-impaired persons, building sanitation such as potable and wastewater systems, as well various structural loads of the building itself and external loads including snow, wind, rain, and seismic ground movements. A building's occupancy classification influences these and other building code provisions. The *International Fire Code* (IFC) is primarily concerned with the safety of emergency responders, that fire protection systems are properly designed, constructed and maintained, the control of combustible materials and ignition sources, and ensuring processes or uses that represent a fire hazard or a high potential of injury or death, such as the release of hazardous materials, are properly designed, constructed, operated, and maintained.

The factors that govern the classification of a building's use must be carefully considered so that those uses or occupancies having approximately the same combustible content and similar fire hazard characteristics will be classified under the same occupancy heading. Occupancies should be grouped so that fire protection requirements and height and area limitations applicable to the occupancy groups are rational for all building uses within that group.

Every classification must be based on the premise that the uses covered by each will have similar fire hazards and life safety problems and that they share like characteristics. Within any given occupancy group or subgroup, no wide differentiation should exist between the fire hazards of the most hazardous and the least hazardous uses.

The occupancy groups include ten major classifications as follows:

A   Assembly
B   Business
E   Educational
F   Factory-Industrial
H   Hazardous

I	Institutional
M	Mercantile
R	Residential
S	Storage
U	Utility and Miscellaneous

In addition to these major classifications, the occupancy groups of Assembly, Factory-Industrial, Hazardous, Institutional, Residential, and Storage are further divided into subgroups in order to accommodate some variations in the hazards associated with the uses within each group (for example, hotel versus an apartment dwelling in the Residential classification). The fire load characteristics in Factory-Industrial and Storage occupancies vary considerably depending upon the product or process involved and, therefore, these uses are further classified into subgroups of low and moderate hazard, depending upon the potential fire severity.

As more and more buildings are being designed either for a single specialized purpose or as a part of a larger type of building complex, the need for more special code considerations have been recognized. Some examples of these special uses include automobile parking structures, domed stadiums, high-rise buildings, covered mall buildings, airport terminals, and large industrial complexes such as steel mills and assembly plants.

Code Administration and Enforcement

Chapter 1: Introduction to Building and Fire Codes

Chapter 2: Legal Aspects, Permits and Inspections

Introduction to Building and Fire Codes

Building and fire codes are a group of regulations that address the construction, alteration, maintenance, and use of buildings. The fire code is unique because it is the only code that establishes requirements for land on which a building is located. The fire code requirements address buildings, including access roadways and water supplies for firefighting operations. The fire code contains certain limitations on how an individual uses their property, as it grants the fire code official the authority to prohibit locations where certain hazardous materials may be stored. It authorizes the code official to regulate site development and construction of buildings located in an area that can be impacted by a wildland fire.

Commonly called construction codes, the separate volumes include requirements for the design of the structure itself based on the materials used to erect the building and the internal and external forces they may face, such as a seismic ground movement, wind load, heavy roof loading due to snow or rain, or flooding. Construction codes have extensive requirements to limit or control the development and spread of an unwanted fire and to ensure the life safety of the occupants. A major consideration of these codes is the health and safety of the building occupants, including requirements for delivering drinking water to the occupants and removing human and food waste for proper treatment and disposal, as well as heating, cooling, and electrical systems to conserve the required energy for these systems while efficiently perform their function. Building construction codes also establish requirements for protecting the envelope from weather and from fires that may initiate in or spread to a building.

The *International Fire Code* (IFC) is one of several construction codes published by the International Code Council (ICC). A major theme of the IFC is the protection of the public from the various hazards that are present in a variety of building uses and activities, as well as the safe storage and handling of hazardous materials. The IFC establishes the minimum requirements for fire department apparatus and firefighter access to buildings, firefighting water supplies, and provisions that address fire hazards inside and outside of buildings. The IFC references the companion International Codes for the construction of buildings regulated by the *International Building Code* (IBC), and one- and two-family dwellings and townhouses, which are regulated by the *International Residential Code for One- and Two-Family Dwellings* (IRC). It also references a variety of national standards, such as those from the National Fire Protection Association (NFPA) or Underwriters Laboratories (UL), which provide specific installation or product details that might not be included in the codes.

This chapter reviews how the IFC and the other codes published by ICC are developed and the scope of the other companion codes, and the chapter concludes with an extensive review of the IFC scope.

CODE DEVELOPMENT

The technology and esthetics of how buildings are constructed, as well as how goods and materials are manufactured, is constantly evolving. Accordingly, the model International Codes are systematically revised at periodic intervals to keep up with technical changes and to address the improved understanding of hazards. The International Codes are revised and updated through an open development process that invites participation by all stakeholders and affected parties. This process can involve exhaustive research, review, and debate of the technical, operational, and administrative issues.

A code change begins with the submittal of a proposal. Any interested group or individual may submit a code change proposal and participate in the proceedings in which it and all other proposals are considered. Following the publication and distribution of the proposals, an open public hearing is held before a committee of representatives from the industry and government, including code officials, contractors, builders, architects, engineers, and industry professionals with expertise related to the applicable code or portion of the code being considered. This open debate and broad participation before the committee ensures that a consensus of the construction community and those impacted by fire and building codes are involved in the decision-making process. The committee may approve, modify, or disapprove the code change proposal.

The ICC membership present at the hearing has the opportunity to overturn the vote of the committee. Following the published results of the hearing, anyone may submit a written comment proposing to modify or overturn the hearing results. The next public hearing is the Final Action Hearing, in which the merits of code change proposals that received public comment are debated. Though any interested party may offer testimony, only ICC Government members (designated public safety officials of a government jurisdiction responsible for administering and enforcing codes) are permitted to cast votes at the final action hearing. The vote is limited to public safety officials because they have no vested financial interest in the outcome and they legitimately represent the public interest. This important process ensures that the International Codes will reflect the latest technical advances and will address the concerns of those throughout the industry in a fair and equitable manner.

A new edition of the code is published every three years. The IFC and all of the other model codes are developed during a single 18-month cycle of the three year period. Code changes to the other codes in the ICC codes family are reviewed and acted on during the later 18-month period. The code changes and the results of the actions by the Code Development Committee, proponent and the membership are published in the Code Change Proposals and Report of the Hearings monographs.

Code Basics

Three year IFC development cycle:
1. Anyone can submit a code change proposal.
2. Proposals are published and distributed.
3. Open public hearings are held before the committee.
4. Public hearing results are published and distributed.
5. Anyone can submit public comments on hearing results.
6. Public comments are published and distributed.
7. An open public final action hearing is held.
8. Final votes are cast by ICC Government members.
9. A new edition of the IFC is published. ●

FIGURE 1-1 International Codes

Although the codes are updated every three years (See Figure 1-1), they cannot be enforced in a community until that jurisdiction adopts them. The 2009 IFC might be published and ready for use, but if the jurisdiction has adopted only the 2006 IFC, the latter is the legally enforceable code. (See Chapter 2)

THE BUILDING AND FIRE CODES—SCOPE

Each of the International Codes has common features that are consistent across the entire library of construction and property maintenance codes. Each code begins by stating its scope. The scope establishes the range of buildings, facilities, construction, equipment, and systems to which the particular code applies. The scope of the IFC is different from the other construction codes because its provisions apply to storage, activities, and hazards located indoors or outdoors of new or existing buildings. The other construction codes are generally limited to new buildings or buildings undergoing renovations—when they regulate outdoors uses or systems, the requirements are limited to a particular building service such as plumbing or fuel gas systems.

The International Codes contain a cross reference to the applicable codes based on the system or element of a building. The IFC makes extensive references to the requirements in the *International Mechanical Code* (IMC) for refrigeration systems, hazardous exhaust mechanical ventilation systems, and commercial cooking operations. The IFC defers to the IBC construction requirements for new buildings or buildings undergoing renovation. Fire code officials are generally are not well-versed in the structural design of buildings or the design of building mechanical, plumbing, and electrical systems, so the IFC defers to the IBC in these instances. The IBC charges the Building Official as the individual responsible for these and other elements regulated by the IBC, including fire-resistive construction, structural integrity, means of egress, and accessibility. Any new buildings or existing buildings that are altered or changed must comply with the requirements in the IFC and the IBC. [Ref. 102.4]

All of the International Codes refer to nationally recognized standards. Standards establish the minimum requirements or criteria for the acceptance of materials, components, assembled equipment or machinery, and systems. Standards are considered as part of the IFC requirements and must be complied with. If an instance arises where a conflict exists between a code and standards, the provisions in the IFC take precedence. [Ref. 102.7]

Each of the model International Codes contains one or more Appendices. Appendices are not mandatory requirements unless they are specifically adopted by the governmental entity. (See Chapter 2)

International Building Code (IBC)

The provisions of the IBC apply to the construction, alteration, maintenance, use, and occupancy of all buildings and structures except detached one- and two-family dwellings and townhouse and their accessory structures which are covered by the IRC. The IBC prescribes various requirements based on the relative risks and hazards of the intended uses within the building and controls the design accordingly. In addition to structural components and systems, the IBC classifies the occupancy of the building, provides for safe interior finish and means of egress, accessibility for mobility impaired individuals, passive and active fire protection, weather resistance, and interior environments. (See Figure 1-3) These requirements are based on the use and occupancy of the building. The occupancy classification of a building or use is based on its intended function. The IBC prescribes various requirements based on the intended uses within the building and controls the design accordingly.

You Should Know

The IFC applies to new and existing buildings and conditions that are hazardous to life, property, or public welfare inside or outside a building. (See Figure 1-2) Its requirements are correlated to other International Codes. ●

FIGURE 1-2 *2009 International Fire Code*

IBC provisions limit the building's height and area, location on property, means of egress, construction, and degree of fire protection, and the provisions vary greatly among covered malls, high-rise buildings, warehouses, night clubs, schools, apartments, and grocery stores.

International Residential Code (IRC)

The requirements in the IRC apply to the construction, alteration, use, and occupancy of detached one- and two-family dwellings and townhouses. Such buildings are limited to not more than three stories above grade in height, and each dwelling unit must have a separate means of egress. (See Figure 1-4) This construction code includes provisions for structural elements, fire and life safety, indoor air quality, energy conservation, and the building's mechanical, electrical, and plumbing systems. The IRC requires compliance with prescriptive construction provisions or the use of performance design criteria. The provisions in the IFC are applicable to the exterior elements of IRC regulated buildings, including premises identification, fire apparatus access, and water supplies. The IFC administrative, operational, and maintenance provisions are also applicable. [Ref. 102.5]

International Wildland-Urban Interface Code (IWUIC)

The *International Wildland-Urban Interface Code* (IWUIC) sets forth requirements for geographical areas of a governmental jurisdiction where structures and other human development meet or intermingle with wildland or vegetative fuels. (See Figure 1-5) IWUIC requirements are applied to mitigate the risk of life and property loss of a fire from wildland fire exposures and fire exposures from adjacent structures and to limit the potential of a structure fire igniting adjacent wildland areas.

FIGURE 1-3 IBC regulated building

FIGURE 1-4 A single-family dwelling. The requirements for its construction or alteration are set forth in the IRC

The IWUIC accomplishes this by requiring the jurisdiction to legally identify and map its Wildland-Urban Interface areas followed by applying scaled construction requirements to improve the ignition-resistance of buildings and structures. The Code also prescribes requirements for establishing and maintaining defensible space around buildings.

International Mechanical Code (IMC)

The *International Mechanical Code* (IMC) regulates the design, installation, maintenance, and alteration of building mechanical systems that are used to control the environment and related processes. The IMC does not apply to the installation of fuel-gas distribution piping and appliances—these systems are regulated by the provisions in the *International Fuel Gas Code* (IFGC).

The IFC makes extensive reference to the requirements in the IMC for refrigeration systems, commercial kitchen cooking systems, fuel oil piping, and hazardous exhaust ventilation systems used in battery rooms, flammable finishing operations, semiconductor fabrication facilities, and buildings storing and handling hazardous materials. (See Figure 1-6)

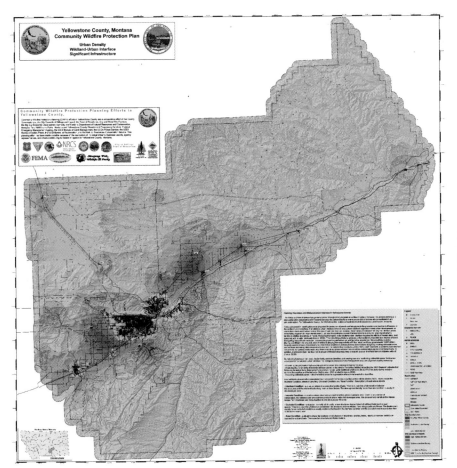

FIGURE 1-6 A commercial kitchen cooking system is regulated by the IMC and IFC. If the appliances are supplied by fuel gas, the system is also regulated by the IFGC

FIGURE 1-5 Wildland-Urban interface map (*Courtesy of Billings (MT) Fire Department*)

International Fuel Gas Code (IFGC)

The *International Fuel Gas Code* (IFGC) regulates the installation of natural gas and liquefied petroleum gas (LP-Gas) systems, fuel gas utilization equipment (appliances), gaseous hydrogen systems, and related accessories. (See Figure 1-7) The scope of the IFGC extends from the utility company's point of delivery to the appliance shutoff valve. Its requirements address pipe sizing and arrangement, approved materials, installation, testing, inspection, operation, and maintenance. The equipment installation requirements include combustion and ventilation air, approved venting, and connections to the fuel gas system. The IFC Chapter 38 requirements references the IFGC requirements for LP-Gas.

International Property Maintenance Code (IPMC)

The *International Property Maintenance Code* (IPMC) establishes minimum regulations for the maintenance of property. Its purpose is to adequately protect the public safety, health, and general welfare of individuals who may be affected by the continuous occupancy of buildings and the maintenance of structures and premises. Existing structures and premises that do not comply with these provisions must be altered or repaired to provide a minimum level of health and safety. The provisions of the IPMC apply to all existing residential and nonresidential structures. It includes minimum requirements for light, ventilation, space, heating, sanitation, protection from the elements, fire and life safety, and for safe and sanitary maintenance. (See Figure 1-8)

FIGURE 1-7 Natural gas meters. The piping on the discharge side of each meter is regulated by the IFGC

FIGURE 1-8 A leaking sewage system would require repair under the provisions in the IPMC

INTERNATIONAL FIRE CODE

The provisions of the IFC address the hazards of fire and explosion arising from the storage, handling, or use of materials, structures, devices, and conditions that are fire hazards or that are hazardous to life, property, or public welfare in any occupancy, structures, or premises. The IFC requirements address the design, construction, installation, testing and maintenance, or removal of fire protection systems, including automatic sprinkler systems and fire alarm and detection systems. Conditions that can affect the safety of firefighters and emergency responders during the emergency operations are also regulated by the IFC. (See Figure 1-9) **[Ref. 101.2]**

The IFC, like other International Codes, is arranged and organized to follow sequential steps that generally occur during a plan review or inspection. The IFC is divided into eight different parts:

Chapters	Subjects
1 – 2	Administration and Definitions
3 – 4	General Safety Requirements
5 – 10	Building and Site Requirements
11 – 26 and 45	Special Processes and Uses
27 – 44	Hazardous Materials
46	Construction Requirements for Existing Buildings
47	Reference Standards
Appendices A-J	Appendices

The IFC requirements for fire-resistive construction, interior finish, fire protection systems, and means of egress are directly correlated to the requirements in the following IBC chapters:

FIGURE 1-9 Part of the IFC scope is the safety of first responders operating at the scene of an emergency

Chapter	Subject
7	Fire-Resistance-Rated Construction
8	Interior Finish, Decorative Materials and Furnishings
9	Fire Protection Systems
10	Means of Egress

Applicability of the IFC

Given its broad scope, the requirements in the IFC are applicable to the design and construction of new buildings or when a change of use or occupancy occurs. Its requirements are intended to be used in conjunction with the requirements of the IBC. The IFC also establishes requirements for the operations and maintenance of a building, processes, or systems located indoors or outdoors. [Ref. 102.2]

Construction and design requirements in the IFC are applicable to:

1. Structures, facilities, and conditions that arise after the Code is adopted.
2. Existing structures, facilities, and conditions not legally in existence at the time the Code is adopted.
3. Existing structures, facilities, and conditions when required by the provisions in IFC Chapter 46.
4. Existing structures, facilities, and conditions which, in the opinion of the fire code official, constitute a distinct hazard to life or property. [Ref. 102.1]

The requirements of the IFC are applicable for any conditions that arise after the date the code is adopted and for any new facilities or structures that are constructed. If a developer wishes to construct a new apartment community and the jurisdiction has adopted the 2009 family of International Codes, then the site and building would be required to comply with the 2009 IFC.

During an inspection, it is common to find a condition that is a violation of the IFC. If the violation is identified during an inspection and the jurisdiction has adopted a fire code, item 2 of Section 102.1 requires that the violation be corrected using the requirements in effect at the time the violation was observed. As an example, a fire inspector performing an inspection of a restaurant and night club finds that portable outdoor gas-fired heating appliances are being used inside of a canopy-covered area of an outdoor balcony. (See Figure 1-10) The jurisdiction has adopted the 2009 IFC. In this case, the inspector would cite the owner of violating Section 603.4.2.1.1, because it is violation of this section to use this type of heater inside of a canopy.

Item 4 of Section 102.1 grants the fire code official the authority to retroactively enforce any provision in the 2009 IFC when, in the opinion of the code official, a hazard is considered to be *distinct*. Declaring a building, process, or use as being a distinct hazard requires a great deal of consideration because once such a declaration has been made, the requirements must be equally enforced on all other similar buildings, processes, or uses. Such a declaration commonly results in legal

FIGURE 1-10 Two portable outdoor gas-fired heaters used inside a building. Based on the requirement in Section 102.1, item 2, a fire inspector would cite this installation as a violation of the requirements in Section 603.4.2.1.1

challenges by the plaintiffs, due to the costs associated with the retroactive construction and renovating systems or features to comply with the IFC. In those instances where a fire code official considers a building, process, or use a distinct hazard, such a decision should be discussed with the jurisdiction's attorney before making the declaration. In many cases, the retroactive application of a particular code requirement should consider the time required to obtain construction permits and to perform inspections, because immediate compliance may be difficult, if not impossible.

Retroactive application of the IFC

Chapter 46 of the IFC stipulates requirements that are applicable to all existing buildings and facilities. These requirements specify the minimum fire safety requirements for existing buildings, such as the protection of vertical openings in shafts. Opening protection in shafts is required to limit the vertical spread of smoke. In occupancies that represent a high life safety risk, such as an elementary school where the children must be under supervision [educational (Group E)], or a hospital, where patients may not be capable of effecting self-rescue [institutional (Group I)], the IFC requires the retroactive installation of a fire alarm and detection system to ensure that an unwanted fire is detected early in its growth. Other retroactive requirements address means of egress components, address numbers or letters on buildings, interior finish, or the installation of automatic sprinkler systems in Group I-2 occupancies.

Change of use or occupancy

The building official will issue a Certificate of Occupancy when a building is approved for occupancy. Every building constructed or renovated under the IBC receives a Certificate of Occupancy. (See Figure 1-11) In addition to the building's address, its building permit number, and the

FIGURE 1-11 New building construction or renovation must comply with the requirements of the IFC and IBC

adopted edition of the IBC used as the basis for its construction, the certificate of occupancy documents:

- The use and occupancy of the building,
- The building's construction type,
- The design occupant load,
- If an automatic sprinkler system is provided, whether the system is required,
- The permitted design live load if a floor is designed to support a live load greater than 50 pounds/square foot and,
- Any special conditions or stipulations issued by the building official.

Fire code officials commonly use the certificate of occupancy during building inspections to verify the building's occupancy or use has not changed. If a building changes occupancy, for example, if a grocery store [mercantile (Group M occupancy)] is converted to a church with an occupant load of 550 [assembly (Group A-3 occupancy)] without a review and approval of the jurisdiction, such an act can result in the fire code official issuing a stop use order. This occupancy change could require the building construction to be upgraded to increase its fire-resistance, providing additional means of egress components, installation of automatic sprinkler manual fire alarm systems, or an upgrade of the building's plumbing system. If the building is not designed for a particular use, it can potentially place the occupants in danger.

The IFC generally prohibits a change of occupancy or building use unless the change is done in conformance with the requirements of IBC. The IFC allows changes in the use of a building provided it does not change the overall use or character of a building. When approved by the fire code official, a building's use can change without conforming to all the requirements of the IBC, provided the new or proposed use is a less hazardous fire risk or life risk. For example, consider a Group S-1 storage occupancy storing rolled carpet. Rolled carpet has a very high heat release rate and requires a specialized automatic sprinkler system design. If the rolled carpet were removed and the warehouse were used for the storing pallet loads of bottled water, the building could be reclassified as a Group S-2 occupancy, because the stored commodity has a much lower heat release rate. **[Ref. 102.3]**

Historic buildings

In 1966, the U.S. Department of the Interior was assigned the responsibility of ensuring historic buildings were preserved under the National Historic Preservation Act. The legislation required each state to establish a historical building preservation office. As a result of this act, many communities also enacted their own local historic building preservation laws.

Historic buildings generally must be maintained in their original condition. Historic buildings may lack fire safety features normally required for new buildings having the same occupancy classification. (See Figure 1-12) These buildings also may not comply with means of egress requirements because they were constructed prior to the development of fire and life safety design regulations in model codes and standards.

FIGURE 1-12 Historic buildings must comply with a fire protection plan approved by the fire code official

Unless the building is a distinct hazard, the IFC requires that historic structures be provided with fire protection and life safety features based on an approved fire protection plan. The criteria for developing a fire protection plan is contained in NFPA Standard 914, *Code for Fire Protection in Historic Structures*. In some cases the fire protection plan may need to be prepared as a performance based design. In these instances, the design should be prepared based on the requirements in ICC *Performance Code® for Buildings and Facilities*. [Ref. 102.6]

Referenced codes and standards

The design, construction, testing, and maintenance of a variety of systems or components is required by the IFC to comply with various technical standards. The IFC adopts over 300 different standards by reference in Chapter 47. Standards are formal documents that establishes consistent and uniform technical or engineering criteria, methods, and practices. When designing an automatic sprinkler system for the protection of a building, the IFC requires that it be designed in accordance with one of the three NFPA standards that govern the design of these systems.

The IFC also requires the evaluation of certain materials or components to be performed using standard test methods, which are definitive procedures for evaluating a product or component. The classification of a liquid as being either flammable or combustible is required by the IFC to be tested in conformance with one of four American Society of Testing and Materials (ASTM) standard tests to measure its boiling point and closed cup flash point temperatures. Compliance with the adopted technical standards or test methods is a requirement of complying with the IFC. (See Figure 1-13) [Ref. 102.7]

FIGURE 1-13 The design of these petroleum storage tanks was required by the fire code official to comply with one of the atmospheric storage tank design and construction standards developed by the American Petroleum Institute

Legal Aspects, Permits and Inspections

Fire codes intend to protect the health and safety of the public and emergency responders by establishing minimum requirements for the prevention of fires and explosions. To be effective, a fire code and its referenced standards must be adopted by a government jurisdiction and enforced by qualified officials appointed by the governing authority. Chapter 2 explains the process of adopting, amending, and administering the *International Fire Code*.

CODE ADOPTION

The IFC and the other International Codes are commonly referred to as *model* codes. Model codes are nationally recognized regulations that address the design, construction, operations, and maintenance of buildings, uses, and hazards that are maintained and updated through an open and accessible code development process. The code development process relies on the participation of interested parties and individuals with experience in the areas of design, construction, manufacturing, logistics, and code administration. The model codes are updated on three-year cycles to recognize new and developing materials, technologies, and methods of construction. Changes to the model codes are commonly in response to natural or technological disasters or in response to acts of terrorism that result in human injury or death or in the destruction of property.

Adoption of the IFC

A governmental jurisdiction must legally adopt the IFC for it to have the power of law. This is normally accomplished by an adopting ordinance that references the title and edition of the IFC. (See Figure 2-1) Generally the legislation also includes the purpose, scope, and its effective date. As part of the adopting legislation, the jurisdiction also provides information for insertion into code text including the name of the jurisdiction, limits for fines or imprisonment as a result of an individual or company violating the fire code, adopted appendices, permit fees, and establish land use limits for the aboveground storage of certain hazardous materials. **[Ref. 101.1, 101.2.1, 111.4, and 113.2]**

Amending the IFC

Adoption of a model fire code offers the jurisdiction with consistency and uniformity in performing plan reviews and inspections, as well as correlation among the adopted construction codes across different jurisdictional boundaries. Such uniformity benefits design professionals, builders, and the regulated industries. While states, counties, and local governments and similar jurisdictions do not have the resources to develop and maintain their own comprehensive fire protection and prevention codes, some jurisdictions have the ability to modify the model fire code through amendments placed in the adopting ordinance. Excessive local amendments to adopted model codes are contrary to the goals of consistency and can offset the advantages and legal defensibility of nationally recognized standards. Some states prohibit amending any construction codes, and the local jurisdiction must enforce what the state legislative or authorized administrative body adopts.

While the model codes anticipate location and climate differences, their amendments are generally influenced by unique characteristics and conditions such as engine or ladder company response times or staffing, the available water supply, community topography, or by land use limitations. There also may be considerations of community politics, customs

ORDINANCE NO. 2009-0009
ADOPTION OF THE 2009 INTERNATIONAL FIRE CODE; ESTABLISHING PERMITS FEES, FINES; DECLARATION OF FIRE CODE VIOLATIONS AS A CRIMINAL VIOLATION OF THE CITY CODE; ESTABLISHING LAND USE BOUNDARIES FOR CERTAIN FLAMMABLE HAZARDOUS MATERIALS

An ordinance the City adopting the 2009 edition of the International Fire Code, regulating and governing the safeguarding of life and property from fire and explosion hazards arising from the storage, handling and use of hazardous substances, materials and devices, and from conditions hazardous to life or property in the occupancy of buildings and premises in the City; providing for the issuance of permits and collection of fees therefore; repealing Ordinance No. 2006-006 of the City and all other ordinances and parts of the ordinances in conflict therewith.

The City Council does ordain as follows:

Section 1. That a certain document, three (3) copies of which are on file in the office of the City Clerk of the City, being marked and designated as the International Fire Code, 2009 edition, including Appendix Chapters A-J (see International Fire Code Section 101.2.1, 2009 edition), as published by the International Code Council, be and is hereby adopted as the Fire Code of the City, in the State of Texas regulating and governing the safeguarding of life and property from fire and explosion hazards arising from the storage, handling and use of hazardous substances, materials and devices, and from conditions hazardous to life or property in the occupancy of buildings and premises as herein provided; providing for the issuance of permits and collection of fees therefore; and each and all of the regulations, provisions, penalties, conditions and terms of said Fire Code on file in the office of the City are hereby referred to, adopted, and made a part hereof, as if fully set out in this ordinance, with the additions, insertions, deletions and changes, if any, prescribed in Section 2 of this ordinance.

Section 2. That the following sections are hereby revised:
Section 101.1. Insert: [NAME OF JURISDICTION]
Section 109.3. Insert: [OFFENSE, DOLLAR AMOUNT, NUMBER OF DAYS]
Section 111.4. Insert: [DOLLAR AMOUNT IN TWO LOCATIONS]

Section 3. That the geographic limits referred to in certain sections of the 2009 International Fire Code are hereby established as follows:

Section 3204.3.1.1 geographic limits in which the storage of flammable cryogenic fluids in stationary containers is prohibited: [JURISDICTION TO SPECIFY]

Section 3404.2.9.5.1 geographic limits in which the storage of Class I and Class II liquids in above-ground tanks outside of buildings is prohibited: [JURISDICTION TO SPECIFY]

Section 3406.2.4.4 geographic limits in which the storage of Class I and Class II liquids in above-ground tanks is prohibited: [JURISDICTION TO SPECIFY]

Section 3804.2 geographic limits in which the storage of liquefied petroleum gas is restricted for the protection of heavily populated or congested areas: [JURISDICTION TO SPECIFY]

FIGURE 2-1 Sample ordinance adopting the IFC

or traditions, regional practices, and other local or state laws. Lengthy amendments to the IFC are not uncommon. Amendments must be legally instituted through the adopting ordinance and laws of the jurisdiction. Otherwise, actions to enforce requirements not contained in the IFC or to waive requirements in the model code will not be in accordance with the law.

Appendices

The appendices are developed in the same manner as the body of the model IFC. However, appendices commonly are judged as being outside the scope and purpose of the model code. Appendices offer supplemental information, alternative methods, or recommended procedures. The information may serve as a guideline or as an example of a recommended practice. In instances where an appendix serves as a source of information, such as Appendix G, which offers equivalent weight and volume values for cryogenic fluids, the jurisdiction is not required to adopt the code text, but the jurisdiction can use the technical content for plan reviews and inspections. Appendix chapters or portions of the chapter that gain acceptance over time are sometimes moved into the main body of the model code. [Ref. 101.2.1]

Local and state laws

The requirements in the IFC are not meant to nullify any local, state, or federal law, and in some instances, those provisions will supersede the requirements in the fire code. For example, Exception 3 of Section 2701.1 exempts off-site transportation of hazardous materials from the IFC when the transportation is in accordance with the U.S. Department of Transportation (DOT) requirements. The Hazardous Materials Transportation Act pre-empts the IFC because DOT has authority over the packaging and transportation of hazardous materials, regardless of the transportation mode. (See Figure 2-2) If a fire code official attempts to regulate hazardous materials transportation, DOT can and has obtained a federal injunction prohibiting the jurisdiction from enforcing IFC requirements.

Careful consideration should be given to the correlation of IFC with other codes and laws of the jurisdiction during the adoption process. Zoning ordinances may have more restrictive land use limitations than those set forth in the IFC. Other local laws that may impact the IFC include

FIGURE 2-2 Off-site transportation of hazardous materials is regulated by the U.S. DOT and their requirements supersede those in the IFC

FIGURE 2-3 Sample Engineer's seal

those regulating public streets, traffic calming, storm water management, fire hydrant locations and spacing, or fire protection water supplies from water wells. State laws may supersede local codes related to certain hazardous materials used as fuel gases, underground storage tanks, and licensing of contractors who design and install fire protection systems. State law also dictates when a registered design professional, including a registered professional engineer or licensed architect, is required to seal design drawings and specifications, and state law sets the licensing requirements for these design professionals. (See Figure 2-3) [Ref. 102.11]

AUTHORITY

The IFC establishes Department of Fire Prevention, commonly referred to as a Fire Prevention Bureau or Office of the Fire Marshal, and designates the individual in charge of the implementation, administration, and enforcement of the code as the fire code official. The appointing authority of the jurisdiction appoints the fire code official. The fire code official is authorized to designate individuals as deputy fire code officials who may perform plan reviews and inspections as well as other technical and administrative staff. The position of a fire code official demands skills, knowledge, and abilities to not only fulfill the duties but to maintain and elevate the credibility of the Fire Prevention Bureau and the Fire Department in the eyes of the public. [Ref. 103.1]

Authority and duties of the fire code official

The IFC charges the fire code official with enforcing the provisions of the fire code and assigns broad authority and discretion to do so. With discretion comes the responsibility to make decisions in keeping with the intent of the IFC. Conversely, the fire code official has no authority to require more than the code stipulates. The IFC authorizes the fire code official to develop policies, procedures, and regulations to clarify the application of the provisions. These policies, procedures, and regulations must be within the spirit and intent of the code, and often they are used to explain how the code official intends to enforce the code. [Ref. 104.1]

To effectively perform the prescribed duties, the fire code official must have an understanding of the legal aspects of code enforcement. While given broad authority for enforcement, including but not limited to issuing a Stop Use order or disconnection of building utilities, the fire code official must also recognize the rights of due process afforded to the public. Equally important to the Fire Prevention Bureau in securing safe buildings, facilities, and uses for the community and its emergency responders is to build the public trust through communication, respect, and fairness so that the organization is viewed as a resource rather than an adversary.

Technical assistance

In certain instances, a permit applicant will submit design drawings and specifications to the Fire Prevention Division that involve a building or system that is complicated or technically challenging. This is

You Should Know

"The fire code official is hereby authorized to enforce the provisions of this code and shall have the authority to render interpretations of this code, and to adopt policies, procedures, rules and regulations in order to clarify the application of its provisions." International Fire Code Section 104.1. ●

not uncommon when dealing with buildings or processes that store, handle, or use hazardous materials, the design of automatic sprinkler and standpipe systems for high-rise buildings, or specialized automatic sprinkler systems protecting high-piled combustible storage. To ensure the commissioned design complies with the requirements of the IFC and its adopted standards, the fire code allows the fire code official to obtain technical assistance. (See Figure 2-4) The fire code official is authorized to use a third-party to review the design drawings and specifications to verify that the design complies with the IFC. The cost of the review and any reports is the responsibility of the permit applicant. The individual performing the review and prepares the report and opinion must be approved by the fire code official. A fire protection engineer or chemical engineer may be requested by the jurisdiction to perform the review and to provide technical guidance to the fire code official. The responsibility for the final approval of the permit rests with the fire code official. [Ref. 104.7.2]

Alternative materials and methods

The IFC is specific in its intention not to exclude the use of any material or method of construction not specifically prescribed by the code, subject to the approval of the fire code official. Given the increasing pace at which technology advances, new and innovative materials and construction methods are being constantly introduced into the market. The fire code official has an obligation, as instructed by the code, to approve such alternatives when it is demonstrated by the design professional or permit applicant that the proposed material or construction method offers equivalent quality, strength, effectiveness, fire resistance, durability, and

FIGURE 2-4 The fire code official can utilize the Technical Assistance provision to ensure that the design and construction of difficult or complicated industrial processes such as this chemical blending operation meets the fire code requirements

safety when compared to the IFC requirements. In many instances, the alternative material or method will exceed the requirements of the fire code and provide a greater level of safety. When evaluating an alternative material or method, it is important to ensure that the proposed method or material meets the intent of the applicable IFC provisions.

One method available to assist code officials reviewing alternative materials and methods are International Code Council's Evaluation Service (ICC-ES) reports. (See Figure 2-5) ICC-ES reports are a resource that is available to code officials to verify that the performance of a system, construction method, or component equal the code requirements. In the absence of ICC-ES reports or sufficient data or documentation, the IFC authorizes the fire code official to require a test of the system, component, or construction method to verify it meets the requirements of the IFC and its adopted standards. Testing must be performed by an approved agency and the test method requires the fire code official approval. In instances where the fire code official does not feel qualified to make a determination as to which test method(s) to use or to review the findings of a test report, the code official can chose to request technical assistance. [Ref. 104.9]

Authority at fires and other emergencies

The scope of the IFC includes conditions affecting the safety of firefighters and emergency responders during emergency operations. The IFC grants the fire code official, the fire chief, or the incident commander at the scene of a fire or other emergency to control and direct the incident scene for the protection of life and property or take any other actions that are necessary in the reasonable performance of duty. (See Figure 2-6) This can include disconnecting building utilities or limiting or prohibiting the movement of people and vehicles that are not authorized at the incident scene. These requirements allow those in charge to deploy firefighting resources based on the size and magnitude of the incident. Consider a large commercial building fire. It may be necessary to lay water supply lines from hydrants located one to two blocks from the fire scene. The IFC requirements grant authorized fire department officials the ability to block roads and streets and to limit the areas in which the public or media have access. [Ref. 104.11]

A fire or medical emergency can emotionally impact family members and friends. In some cases, people can become extremely agitated, especially if it's a close family member, friend, or family pet. People can lose focus that the emergency responders are trained professionals and that the methods for controlling fires or patient treatment may not be understood by those observing the emergency. To ensure the care of patients is not compromised and to protect the safety of emergency responders treating the individual or managing the emergency, the IFC allows the incident commander, fire chief, or fire code official to barricade the scene. The IFC also grants emergency personnel the authority to have individuals detained who may be obstructing operations if they disobey a lawful command from fire or police department personnel. (See Figure 2-7) [Ref. 104.11.1, 104.11.2]

ESR-4802

Issued March 1, 2008

This report is subject to re-examination in one year.

ICC Evaluation Service, Inc.

www.icc-es.org

Business/Regional Office ■ 5360 Workman Mill Road, Whittier, California 90601 ■ (562) 699-0543
Regional Office ■ 900 Montclair Road, Suite A, Birmingham, Alabama 35213 ■ (205) 599-9800
Regional Office ■ 4051 West Flossmoor Road, Country Club Hills, Illinois 60478 ■ (708) 799-2305

DIVISION: 07—THERMAL AND MOISTURE PROTECTION
Section: 07410—Metal Roof and Wall Panels

REPORT HOLDER:

ACME CUSTOM-BILT PANELS
52380 FLOWER STREET
CHICO, MONTANA 43820
(808) 664-1512
www.custombiltpanels.com

EVALUATION SUBJECT:

CUSTOM-BILT STANDING SEAM METAL ROOF PANELS: CB-150

1.0 EVALUATION SCOPE

Compliance with the following codes:

■ 2006 *International Building Code®* (IBC)

■ 2006 *International Residential Code®* (IRC)

Properties evaluated:

■ Weather resistance

■ Fire classification

■ Wind uplift resistance

2.0 USES

Custom-Bilt Standing Seam Metal Roof Panels are steel panels complying with IBC Section 1507.4 and IRC Section R905.10. The panels are recognized for use as Class A roof coverings when installed in accordance with this report.

3.0 DESCRIPTION

3.1 Roofing Panels:

Custom-Bilt standing seam roof panels are fabricated in steel and are available in the CB-150 and SL-1750 profiles. The panels are roll-formed at the jobsite to provide the standing seams between panels. See Figures 1 and 3 for panel profiles.

The standing seam roof panels are roll-formed from minimum No. 24 gage [0.024 inch thick (0.61 mm)] cold-formed sheet steel. The steel conforms to ASTM A 792, with an aluminum-zinc alloy coating designation of AZ50.

3.2 Decking:

Solid or closely fitted decking must be minimum $^{15}/_{32}$-inch-thick (11.9 mm) wood structural panel or lumber sheathing, complying with IBC Section 2304.7.2 or IRC Section R803, as applicable.

4.0 INSTALLATION

4.1 General:

Installation of the Custom-Bilt Standing Seam Roof Panels must be in accordance with this report, Section 1507.4 of the IBC or Section R905.10 of the IRC, and the manufacturer's

published installation instructions. The manufacturer's installation instructions must be available at the jobsite at all times during installation.

The roof panels must be installed on solid or closely fitted decking, as specified in Section 3.2. Accessories such as gutters, drip angles, fascias, ridge caps, window or gable trim, valley and hip flashings, etc., are fabricated to suit each job condition. Details must be submitted to the code official for each installation.

4.2 Roof Panel Installation:

4.2.1 CB-150: The CB-150 roof panels are installed on roofs having a minimum slope of 2:12 (17 percent). The roof panels are installed over the optional underlayment and secured to the sheathing with the panel clip. The clips are located at each panel rib side lap spaced 6 inches (152 mm) from all ends and at a maximum of 4 feet (1.22 m) on center along the length of the rib, and fastened with a minimum of two No. 10 by 1-inch pan head corrosion-resistant screws. The panel ribs are mechanically seamed twice, each pass at 90 degrees, resulting in a double-locking fold.

4.3 Fire Classification:

The steel panels are considered Class A roof coverings in accordance with the exception to IBC Section 1505.2 and IRC Section R902.1.

4.4 Wind Uplift Resistance:

The systems described in Section 3.0 and installed in accordance with Sections 4.1 and 4.2 have an allowable wind uplift resistance of 45 pounds per square foot (2.15 kPa).

5.0 CONDITIONS OF USE

The standing seam metal roof panels described in this report comply with, or are suitable alternatives to what is specified in, those codes listed in Section 1.0 of this report, subject to the following conditions:

5.1 Installation must comply with this report, the applicable code, and the manufacturer's published installation instructions. If there is a conflict between this report and the manufacturer's published installation instructions, this report governs.

5.2 The required design wind loads must be determined for each project. Wind uplift pressure on any roof area must not exceed 45 pounds per square foot (2.15 kPa).

6.0 EVIDENCE SUBMITTED

Data in accordance with the ICC-ES Acceptance Criteria for Metal Roof Coverings (AC166), dated October 2007.

7.0 IDENTIFICATION

Each standing seam metal roof panel is identified with a label bearing the product name, the material type and gage, the Acme Custom-Bilt Panels name and address, and the evaluation report number (ESR-4802).

FIGURE 2-5 ICC-ES report

FIGURE 2-6 The IFC grants the officer in charge of controlling this structure fire complete control of scene which includes limiting access to vehicles and the public

FIGURE 2-7 The fire chief, fire code official, or officer in charge of the incident has the authority in the IFC to secure the emergency scene and order the removal of individuals who obstruct operations

Code Basics

Fire code official duties:

- Enforce the IFC
- Review construction documents and permit applications
- Issue permits, notices and orders
- Conduct inspections
- Maintain records
- Investigate the cause and origin of unwanted fires
- Control the scene of emergencies

Fire code official authority:

- Make interpretations
- Adopt policies, procedures, and regulations
- Approve alternative methods, materials, and modifications ●

PERMITS

A permit is a written authorization issued by a code official that legally authorizes an individual or business to conduct certain businesses, services, or construction in accordance with the requirements of the jurisdiction's adopted codes. The IFC requires the fire code official to issue permits to perform certain hazardous operations or activities and for the construction or alteration of fire protection systems and processes storing and handling hazardous materials or highly flammable materials. Under specified circumstances, the IFC exempts certain hazardous operations and activities and permits limited renovations to fire protection systems—however, any work that is exempt from a permit still must comply with the applicable IFC requirements. [Ref. 105.1 and 105.3]

If the construction, alteration, or operation of the system or process is not performed and maintained in conformance with the IFC requirements in effect when the permit was issued, the fire code official can revoke the permit. Permit revocation essentially is a stop use order and the activities must cease until violations are corrected and compliance is demonstrated to the fire code official. [Ref. 105.3.8]

Operational and construction permits

The IFC authorizes the fire code official to issue permits for certain hazardous operations, which are termed as Operational permits, and permits for the construction or alteration of fire protection systems and equipment or systems designed for the storage and use highly flammable materials or hazardous materials, which are identified as Construction permits. The 2009 IFC requires an Operational permit for 46 hazardous processes or activities that are regulated by the code. Operational permits are issued for a prescribed duration or until the permit is renewed or revoked. The jurisdiction will need to establish a policy or procedure for the time duration permitted for Operational permits.

The IFC requires 14 different Construction permits. In addition to the construction or alteration of fire protection systems, including automatic sprinkler systems, standpipes, and private fire protection fire hydrants and water distribution piping, construction permits are required for temporary membrane structures and tents, and for systems storing and using compressed and liquefied compressed gases, cryogenic fluids, flammable and combustible liquids, and other hazardous materials. Construction permits usually expire when the work has been completed and approved by the code official. [Ref. 105.1.2]

It is very common for a particular address or regulated use to require both a construction permit and an operational permit. (See Figure 2-8) For example, a typical motor vehicle fuel-dispensing station will store, handle, and dispense flammable and combustible liquids. The liquids will be stored in aboveground or underground storage tanks which will deliver these liquids, via a pump and piping network, to motor vehicle fuel dispensers. Company personnel and the general public may have access to the dispensers. In such a case, a construction permit is required for the installation of the tanks, piping, pumps, and dispensers, and an Operational permit is required for the storage, handling, use, and dispensing of flammable and combustible liquids. A construction permit is not required for the maintenance of the installed system. [Ref. 105.1.3]

Construction documents

Construction drawings and specifications must accompany any construction permit application and be of sufficient detail and clarity to verify compliance with the code. In some cases, the code will require a site plan showing all new and existing structures with distances to buildings on the

FIGURE 2-8 This Type 2 permanent explosives magazine requires IFC construction and operational permits issued and approved by the fire code official

same property, building openings, and property lines. The extent of construction documents varies with the complexity and scope of project.

Construction documents are defined by the IFC as *the written, graphic and pictorial documents prepared or assembled for describing the design, location and physical characteristics of the elements of the project necessary for obtaining a permit.* Construction documents must comply with the preparation and submittal requirements in Section 105.4. When required by the jurisdiction or state laws, the construction documents must be prepared by a registered design professional, such as a licensed architect or engineer. The fire code official can waive this requirement when it is demonstrated that the nature of the work does not require the services of a registered design professional. [Ref. 105.2, 105.4.1]

Fire protection system drawings and supporting calculations must be prepared in accordance with the applicable NFPA standards referenced in Chapter 9. These plans must also comply with any specific requirements set forth in Chapter 9, such as fire protection of wooden decks and balconies in wood frame Residential (Group R) buildings or the use of quick-response sprinklers or residential sprinklers in certain Institutional occupancies. The fire code places the responsibility for the preparation of the construction documents on the permit applicant and requires each permit application be complete. (See Figure 2-9) [Ref. 105.4.3]

During large or extended construction projects, the registered design professional may want to use a phased design approach in which construction documents are submitted based on various project milestones. In many buildings the use of these *design-build* construction methods have been proven major cost savings because the building is designed as each of major elements is being constructed. Phased approval of construction documents is allowed by the IFC—however, the designer and the contractor are required to assume any risks associated with improper or incomplete construction methods or installations and that as the project concludes, the fire code official may withhold any approvals until all of the code requirements are satisfied. [Ref. 105.4.4.1]

Permit application

Before a permit can be issued, the owner or an authorized agent must apply in such form and detail as required by the fire code official. When required by the IFC or the fire code official, construction documents also must accompany the permit application. Because the degree and level of information can vary between Operational and Construction permits, the jurisdiction should establish a clear procedure and policy for the minimum information required for each permit that is issued by the Fire Prevention Division.

The fire code official is required by the code to examine each permit application to determine if the proposed activity or construction complies with the applicable code requirements. If the submitted information or construction documents do not comply with the requirements in the fire code or its adopted standards, the fire code official can reject the submittal and must provide a written reason to the applicant for its denial. [Ref. 105.2.4 and 105.2.1]

Phoenix Fire Department
Fire Prevention – Plan Review

Date: _____ Initials: _____

Permit: _____

Plan Review Application—
Applicant to fill in area within BOLD LINES

Check One
☐ 1st **Review**   ☐ 2nd **Review**   ☐ 3rd **Review**   ☐ **Revision**   ☐ **Other**

Kiva #: _____   SDEV/SPAD#: _____   Reviewer: _____

Project Name: _____   **Project Number:** _____

DEVELOPMENT INFORMATION

ADDRESS:	BLDG #:	SUITE/SPACE #:	FLOOR #:	ZIP CODE:

DESCRIPTION OF WORK:

SQ. FT. : _____   **# OF STORIES:** _____   **VALUATION:** _____

APPLICANT: (Contact Person)

☐ Owner/Devel. ☐ Arch. ☐ Engr. ☐ Contractor

FIRM NAME:		
ADDRESS:		
CITY:	STATE:	ZIP:
TELEPHONE:	FAX:	Other:

OWNER INFORMATION: (Business/Owner Name)

CONTACT PERSON:	TELEPHONE:	FAX:
ADDRESS:	CITY:	STATE:	ZIP:

CONTRACTOR INFORMATION: (Business/Owner Name)

CONTACT PERSON:	TELEPHONE:	FAX:
ADDRESS:	CITY:	STATE:	ZIP:
BUSINESS LICENSE #:	STATE TAX #:	STATE LICENSE #:

APPROVED _____ APPR AS NOTED _____ CORR _____ HOLD _____ DATE COMPL _____

Rev. 7/07   I:\FP\FP FORMS\New Service Fees & Applications\Plan Review Application.doc

FIGURE 2-9 Plan review application for a construction permit

In many cases, a plans examiner may find that the application and its supporting documentation are acceptable—however, certain portions or features of the design may need to be modified to meet the intent or letter of the fire code. In such cases, the fire code official can issue

FIGURE 2-10 The fire code official can require an inspection before issuing an operational permit

an approval of the permit but stipulate certain conditions or requirements that must be satisfied. The code official cannot waive or set aside any other provisions of the IFC, the jurisdiction's adopted construction codes, or other laws. **[Ref. 105.3]**

The fire code official is authorized to perform an inspection of any building, process, or systems to be used before the permit is issued. This inspection can be used to establish any operational constraints or limits as well as to determine if any other operational permits are required. (See Figures 2-10 and 2-11) **[Ref. 105.2.2]**

Fees

Jurisdictions generally establish a fee schedule at a level to offset the costs of providing services to the public, including administration, plan reviews, and inspections. Most jurisdictions choose not to charge fees for inspections performed by fire companies—however, if a fire company inspection finds a permit that should be referred to the Fire Prevention Bureau, then fees for these services can be assessed. Permit fees are commonly based on the total value of the construction project or on inspection history for various occupancy classes or equipment. Whatever the method selected, the fire code official should develop an equitable and consistent procedure for assessing and refunding fees related to the permit process. **[Ref. 113]**

INSPECTIONS

Inspections are an important part of confirming and verifying fire code compliance. For new construction or tenant renovations, inspections are performed to confirm that fire protection systems are installed in accordance with the approved design presented in the construction documents and in conformance with the adopted NFPA standards. In

City of Fire - Fire Prevention Division - Construction Permit

PERMIT NUMBER: 2009-100-008

PERMIT DATE: January 25, 2009

DESCRIPTION: Automatic Sprinkler System - New Const.

EXPIRATION DATE: January 25, 2010

ADDRESS: 10301 Sweetwater River Cove

BUILDING/SUITE: Bldg. B

SCOPE OF WORK: Wet-pipe automatic sprinkler system - High Piled Combustible Storage

SCOPE OF WORK: All City of Fire Regulations and the 2009 International Fire Code shall apply. Systems are subject to field inspection, to schedule call (512) 555-4321. This permit shall expire (12) twelve months from the date of issue. This permit is not transferrable.

SCOPE: INSTALLATION OF WET-PIPE AUTOMATIC SPRINKLER SYSTEMS USING EXISTING RISERS FOR THE PROTECTION OF PALLETIZED CLASS IV COMMODITIES (GROUP B PLASTICS) TO 17 FT. DESIGN AREAS 1, 2, 3 & 5 PROTECTED USING CENTRAL MODEL K17-231 (K=16.8) PENDANT SPRINKLERS. DISCHARGE DENSITY: 0.44 GPM/SQ.FT. DESIGN AREA: 3000 SQ.FT. DESIGN BASIS IS FACTORY MUTUAL DATA SHEET 8-9 FOR THE INDICATED DESIGN AREAS. DESIGN AREA 4 IS AN EXTRA HAZARD GROUP 2 DENSITY (NFPA 13 DESIGN CRITERIA) PROTECTING THE FLAMMABLE LIQUID PUMP ROOM (GROUP H2 OCCUPANCY).

SEE THE PLAN REVIEW REQUIREMENT FOR SPECIAL INSPECTION OF PIPE HANGERS & SUPPORTS FOR THE 8-INCH FEED MAINS.

DESIGN AREA 1 DEMAND (BOR): 1528 GPM @ 33.3 PSI (SPRINKLER K=16.8)
DESIGN AREA 2 DEMAND (BOR): 1523 GPM @ 42.6 PSI (SPRINKLER K=16.8)
DESIGN AREA 3 DEMAND (BOR): 1672 GPM @ 49.1 PSI (SPRINKLER K=16.8)
DESIGN AREA 4 DEMAND (BOR): 618 GPM @ 49.7 PSI (SPRINKLER K=8.0)
DESIGN AREA 5 DEMAND (BOR): 1476 GPM @ 32.1 PSI (SPRINKLER K=16.8)

CONTRACTOR NAME: South Texas Fire Protection Engineering

CONTRACTOR ADDRESS: 12520 Split Rail Parkway, Jollyville TX, 78750

RESPONSIBLE MANAGING EMPLOYEE: Odean Puryeyor, NICET D-08911

PERMIT FEE: $1,350

PAYMENT: American Express xxxx xxxxxx xxx xxxx 760

Approved plans shall be available and permits shall be posted on-site at all times during which work authorized hereby is in progress. Progress is defined as the time from which site work begins until the time of final Fire Department approval. Inspections shall not be conducted if approved plans and permits are not on-site. Violation of this requirement is subject to a Stop Use order and civil/criminal sanctions as allowed by code.

FIGURE 2-11 A fire code construction permit

many jurisdictions, a fire inspection is a requirement for licensing of day care and health care occupancies. Inspections to verify compliance with the fire code are also required by many building code officials before a certificate of occupancy is issued. In the case of the IFC, an inspection is required before an operational permit can be issued.

Right of entry

IFC Section 106 establishes provisions that authorize the fire department staff to conduct inspections. Because fire inspections can be required for any property, vehicle, or vessel, permission to perform the inspection must be obtained from the property owner, tenant, or an individual authorized to allow entry onto the property. A property owner can refuse an inspector's entry into a building or onto a site, as this right is protected by the Fourth Amendment to the U.S. Constitution:

> The right of the people to be secure in their persons, houses, papers and effects against unreasonable searches and seizures, shall not be violated, and no Warrants shall be issued, but upon probable cause supported by Oath or affirmation and particularly describing the place to be searched and the persons or things being seized.

In 1967, the U.S. Supreme Court issued a ruling involving a fire inspection in the case of *See versus The City of Seattle* (387 U.S. 541, 87 S.Ct. 1737, 18 L.Ed.2d 943). The appellant (Norman See) sought reversal of his conviction for refusing to allow a representative of the City of Seattle Fire Department to enter and inspect his locked commercial warehouse without a warrant and without probable cause to believe that a violation of any municipal ordinance existed. The inspection was conducted as part of a routine, periodic city-wide canvass to obtain compliance with Seattle's Fire Code. After he refused the inspector access, Mr. See was arrested and charged with violating the fire code. He was convicted and given a suspended fine of $100, despite his claim that authority to inspect provision of the fire code, if interpreted to authorize this warrantless inspection of his warehouse, would violate his rights under the Fourth and Fourteenth Amendments. The Supreme Court agreed and issued its ruling in Mr. See's favor, which has governed how code officials and inspectors must initiate inspections.

To satisfy the requirements of the Fourth Amendment and the rulings of the Supreme Court Justices, fire code officials must respect the requirements of the law and perform certain steps before initiating an inspection on or into a protected area. A protected area is not visible from any public view or areas where privacy is expected. In almost every case where the requirements of the IFC are followed, entry will be granted by the individual responsible for the building or premises so the inspection can be performed (See Figure 2-12):

1. Identify yourself, the basis and reason for inspection. The jurisdiction should issue an official method of identifying code officials. **[Ref. 104.4]**

2. Obtain permission and consent from a responsible individual with the business, building, or site. Note that the courts have ruled against third-party consent to conduct an inspection when it was granted by the child of an owner or an employee. Therefore, it is important to obtain either oral or written consent from an individual who is responsible for the site or building.

FIGURE 2-12 Before performing an inspection the fire code official must obtain consent from the owner or a responsible party

3. If the inspection is for the purpose of verifying compliance with a Construction or Operational permit required by the IFC, inform the individual that this is the basis for inspection.
4. In certain cases, a business will request a copy of the legal basis for an inspection. If asked, either cite or provide a copy of IFC Sections 106 and 104.3.
5. Once consent is granted, the inspection can proceed.
[Ref. 104.3]

If access is denied, the inspector should document the circumstances, time and date of the incident, and contact the jurisdiction's legal counsel for advice on how to proceed.

Liability

Fire code officials that perform code enforcement inspections often are anxious about their potential liability exposure if they make a mistake or overlook a potential problem while performing inspections. News reports describe multi-million dollar settlements, the firefighters' unfamiliarity with the code enforcement process, and an inherent suspicion of the legal profession combine to raise the apprehension level. "Can I get sued?" "Will I be held liable?" "Who will protect me?" are common questions for the inexperienced code official as well as fire suppression personnel who may be assigned to code enforcement.

Any person can sue another for some alleged error or insult. A code official can be held liable or found negligent only if it is established after all the facts have been heard at a trial and a jury renders a verdict. Fortunately, there are many protections built into the American jurisprudence system that make the likelihood of being found liable a remote possibility as long as fire code official or fire inspector is acting within the scope of the authority and training, called a "standard of care." There are protections in the *International Fire Code* itself that limit a code official's or firefighter's liability. The fear of being sued should never interfere with a jurisdiction's commitment to perform code enforcement for the safety of firefighters and the public.

The laws regarding liability are incredibly complex and differ in every state; a detailed discussion is beyond the scope of this book. In some states, government officials are protected from lawsuits arising from any work they perform in their governmental roles (they are said to possess "governmental immunity"), but this principle has been eroded in many states in recent years. If code official or inspector is concerned about liability exposure while performing inspections or code enforcement, the individual should discuss those issues with legal counsel. It might be a city or county attorney, or a professional who is on retainer to provide legal advice and consultation.

In very general terms, to prove someone is liable of negligence, the person bringing the charge (the "complainant" or "plaintiff") must prove the defendant is guilty of four things: duty, breach of duty, proximate cause, and damages.

"Duty" describes conditions where the defendant had a legal obligation to perform a specific task, and often that specific task has to be provided for a certain group or "class" of persons. For example, in many states, firefighters and emergency medical services personnel have a legal obligation—a duty—to report child or elder abuse to law enforcement. These same first responders may not have a duty to report abuse to someone who does not meet the legal definition a "child" or "elder" citizen. The duty or obligation to perform building and fire code enforcement varies by jurisdiction.

"Breach of duty" means the defendant failed to fulfill the obligation. If an inspector has a legal obligation to report electrical hazards to the local electrical inspector and fails to do so, he may be accused of breaching that duty.

"Proximate cause" means that whatever the alleged code violation or oversight might have been, it has to have lead to the problem for which the defendant is accused. Imagine that a firefighter has a duty to inspect sprinkler fire department connections and fails to check one on a commercial inspection. Later, it is determined that the connection had been badly damaged by a delivery truck and the fire department could not use it during a fire. The complainant might argue successfully that had the fire department been able to use the fire department connection, the fire's outcome might have been different. There is a link (a "proximate cause") between the failure to inspect the connection and the inability to use it.

Finally, there must be "damages." Without damage, there is no liability. In the previous example, if the fire was controlled by a single sprinkler and extinguished by hand lines without spreading beyond the area of origin, the plaintiff would be hard pressed to claim there was any damage caused by the broken fire department connection because it did not play a part in the incident.

If the plaintiff cannot prove all four elements (duty, breach of duty, proximate cause, and damages) the defendant likely will not be found guilty of negligence or liability.

The IFC provides another level of protection against liability. The IFC states the *fire code official, officer or employee charged with the enforcement of this code, while acting for the jurisdiction, shall not thereby be rendered liable personally, and is hereby relieved from all personal liability for any damage accruing to persons or property as a result of an act required or permitted in the discharge of official duties.* This protection extends to the fire code official, firefighter, or company officer conducting inspections within the scope of his work. The IFC also requires the jurisdiction employing the fire code official to provide legal defense and to cover the costs of any settlement that might result from a trial [Ref. 103.4.1]

Testing and operation

The fire codes place the responsibilities of the proper installation of any required fire protection, life safety, or hazardous materials storage and use systems in the hands of the registered design professional or the installing contractor. Once these systems are approved by the fire code official, it is the owner's or tenant's responsibility to ensure that they are maintained in accordance with the fire code requirements, including its adopted standards. [Ref. 107.5]

Many of the systems specified by the IBC and IFC are required to be inspected or tested annually. These inspections are the responsibility of the owner or tenant, and most businesses will retain a contractor to perform required inspections and any maintenance. It is the responsibility of the owner or tenant to maintain these records so they can be reviewed by the fire code official. In some jurisdictions, system testing and inspection reports are remitted to the fire code official for review. [Ref. 107.2.1]

During these system inspections and tests, the contractor or fire code officials may identify fire code violations that require correction. (See Figure 2-13) The IFC requires that the necessary corrections be performed and that the fire code official inspect or test the corrected area of concern to verify compliance. [Ref. 107.2.2]

FIGURE 2-13 The fire code official is authorized to perform as many inspections as necessary to verify that systems are properly installed and maintained

Unsafe buildings

If, as a result of a natural or technological emergency, or as a result of the lack of maintenance or changing a building's use, a fire code official finds that a building represents a serious fire or life safety threat, the IFC grants broad authority to require the necessary corrections to bring it into code compliance. The scope of this authority extends to building systems or any system regulated by the fire code. Any actions necessary to satisfy the requirements of the IFC that involve repair or upgrades to building structural components or building demolition require the approval of the building code official. When beginning the process of correcting unsafe buildings or conditions, a written notice to the owner, tenant, or responsible party is required. [Ref. 110.1]

If the fire code official finds that the violation constitutes a hazardous condition that presents an imminent danger to the building occupants, the code official is authorized to require the partial or complete evacuation of a building and can prohibit re-entry into the building. (See Figure 2-14) [Ref. 110.2]

Stop work order

In some instances, individuals or businesses will perform work that is regulated by the IFC without obtaining the required Operational or Construction permits. Under certain circumstances, the continued work or operation of equipment represents dangerous or unsafe conditions to employees or personnel who may have access to the site or building. In these cases, the fire code authorizes the fire code official the authority to issue a Stop Work Order. (See Figure 2-15) When a fire code official issues such an order, it must be in writing, must explain the basis for stopping the work, and must state the conditions under which the cited work is allowed to resume. Unless otherwise stipulated by the fire code official, all Stop Work Orders are an immediate compliance order—when issued,

FIGURE 2-14 If a building is found to be an imminent hazard the fire code official can require its evacuation and prohibit re-entry *(Courtesy of New Orleans [LA] Fire Department—Photo Unit)*

FIGURE 2-15 This gasoline underground storage tank was not adequately secured and floated aboveground after extended rains saturated the soil. Because of the fire and unauthorized release of hazardous materials threat, the fire code official issued orders for the removal of the stored flammable liquid and the damaged storage tank

the activity affected by the legal order must stop immediately. Failing to comply with the Stop Work Order is generally treated by most jurisdictions as a criminal violation and can be subject to fines. [Ref. 111]

BOARD OF APPEALS

The IFC administrative provisions create authority and duties for the fire code official but intends that actions in enforcing the code be reasonable while protecting due process rights for the regulated community. Under the provisions of the IFC, a property or business owner has the right to legally challenge the code interpretations of a fire code official or inspector. Any person or organization that has a material interest in the fire code official's decision may apply for a hearing or review to the Board of Appeals. (See Figure 2-16) The jurisdiction's governing body, such as its board of directors or city council, appoints the board members, who are qualified by experience and training to hear and rule on interpretations issued by the fire code official.

The IFC limits the basis for appeals to matters pertaining to code requirements. The appellant must claim the fire code official has erred in interpreting the code or has wrongly applied a code section. Another common basis for an appeal is to request the Board of Appeals to consider alternative methods and materials as being equivalent to the code requirements. The IFC does not allow the filing of an appeal to seek a variance or waiver and does not grant the Board of Appeals the authority to waive code requirements. [Ref. 108, Appendix A]

City of Phoenix

PHOENIX FIRE DEPARTMENT
Fire Prevention Section
150 South 12th Street
Phoenix Arizona 85034-2301
(602) 262-6771 FAX: (602) 271-9243

Petition of Appeal to the Fire Marshal

All appeals shall be detailed on this form. Supporting data may be attached and submitted if desired however, all entries and statements on this form shall be complete. Incomplete forms will not be accepted.

INTERNAL USE:

Log Number:	Date Logged Out:	Date Logged In:
Case/KIVA Number:	Hearing Date:	Hearing Time:
Engineer/FPS Familiar with Project:	Occupancy Type:	Compliance Date:

Business/Occupancy Name:	Address:
Business Owner's or Corporate Agent's Name:	Mailing Address:
Tenant's Name:	Mailing Address:
Appellant's Name:	Mailing Address:

This appeal applies to (Check one):

☐ A project in the plans review stage. Building Safety Log No. _____

☐ An alleged Fire Code violation.

An appeal is hereby made to the Fire Marshal for a deviation from Section _____ of the Phoenix Fire Code.
Briefly state the requirements being appealed.

State in detail what is proposed in lieu of literal compliance with the Fire Code:

Appellant's Signature:	Title:	Phone Number:
Building Owner's Signature:		Building Owner's Phone Number:

INTERNAL USE:

Decision of the Fire Marshal

☐ Approved   ☐ Approved with Stipulations
☐ Denied   ☐ See Attachment

Fire Department Official:	Date:

DISTRIBUTION: WHITE – Appeals File   YELLOW – Appellant   BLUE – Fire Prevention   *WEB – 3 Completed & Signed Forms Required for Submission

91-48D Rev. 3/05
61582253060-CP

FIGURE 2-16 Application to the Board of Appeals

Courts have ruled that permit and inspection fee requirements cannot be waived because they are costs associated with an inspection program. However, a permit applicant still has the right to appeal.

General Safety Requirements

Chapter 3: General Precautions Against Fire

Chapter 4: Emergency Planning and Preparedness

General Precautions Against Fire

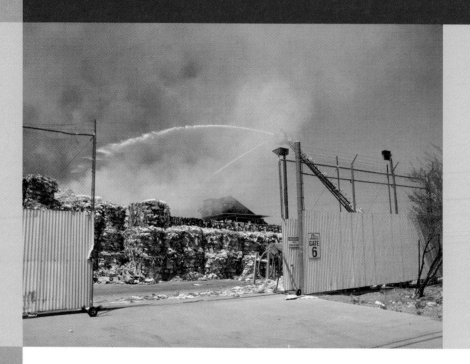

The general safety requirements in the IFC were developed to control a wide variety of fire safety concerns that may not need additional clarification or the level of detail that might be found in other chapters.

Chapter 3 covers combustible waste materials (such as wood, paper, and plastics) and sources of ignition. Ignition sources include mechanical, chemical, electrical, or optical energy. The chapter also addresses topics such as fire safety issues related to vacant premises and fueled equipment. Vacant premises can be a major fire hazard to communities because if they are not adequately secured, they may be used for criminal activity or as illegal and substandard housing. "Fueled equipment" includes motorcycles, mopeds, lawn-care equipment, and portable cooking equipment. Fueled equipment are found in a variety of buildings and work sites, and they represent another fire hazard because of the fuels they use and their common indoor use.

COMBUSTIBLE MATERIALS

Combustible materials are natural or synthetic materials that can be ignited and support combustion. Combustible materials in the context of this chapter and IFC Chapter 3 are not combustible metals or flammable solids—these are hazardous materials that are regulated by other provisions in the fire code. Materials regulated by IFC Chapter 3 generally are organic materials such as sawn wood, dimensional lumber, waste paper or cardboard, and baled cotton or paper. Synthetic materials may include plastics, fabrics, or composite materials. Combustible materials are always solids and will have varying sizes and densities. The smaller the area of a combustible material and the lighter its density, the more easily it is ignited. The orientation of the combustible material, the strength of the ignition source, and a variety of other variables can influence the ignition of combustible materials.

The fire code recognizes that combustible materials are an important part of the businesses and industries. The combustible material requirements in IFC Chapter 3 address the orderly storage of these materials, locating the materials away from ignition sources, and, if the storage is indoor, separating the combustible materials from means of egress components and concealed spaces where they could accelerate the rate an unwanted fire grows and spreads. Orderly storage can slow the rate of fire spread, which benefits firefighters in the event the materials are ignited. (See Figure 3-1) **[Ref. 315]**

While it is not within the scope of this chapter, fire code officials should understand that storage of many combustible materials over 12 feet in height inside of buildings introduces the potential for a fire that will exhibit a much faster growth rate when compared to the same materials stored at or near the floor level. Such storage can be found in many warehouses and mercantile occupancies and is required to comply with the requirements in IFC Chapter 23. Chapter 14 introduces the reader to the hazards of high-piled combustible storage.

When combustible materials become "waste," the IFC takes a more aggressive approach: the materials must be removed and disposed of in a controlled manner. For most combustible wastes, the IFC requires that it be placed in noncombustible waste containers or plastic containers formulated from chemicals that reduce the amount of heat it releases if ignited. When materials are placed in bulk trash receptacles (dumpsters), the fire code requires they be located at least 5 feet from combustible construction, wall openings, and combustible roof eaves. (See Figure 3-2) Because of land use limitations, it is very common to place dumpsters inside of buildings. In such instances, the room housing the dumpster is required to be protected by an automatic sprinkler system. Sprinkler protection is not required when the dumpster is located in a building constructed of noncombustible, fire-resistive materials. **[Ref. 304.3]**

FIGURE 3-1 Storage of combustible materials beneath the egress stairs is a violation of the IFC. The concern is if the materials are ignited, this means of egress component is no longer available for the safe evacuation of the occupants

FIGURE 3-2 The IFC requires separation of outdoor dumpsters from buildings to limit the likelihood of the dumpster igniting an exposure building

IGNITION SOURCES

Controls for ignition sources are dictated in several chapters of the IFC, including specific requirements for electrical equipment and hot work involving brazing, oxygen-acetylene cutting, and welding. IFC Chapter 3 general requirements address separating uses and activities involving potential sources of open flames from combustible materials. The provisions require adequate separation between open flames and combustible materials, open-flame warning devices such as road flares, and negligent burning of combustible vegetation and materials. Cooking, decoration, theatrical, or construction activities are regulated elsewhere in Chapter 3. [Ref. 305]

OPEN FLAMES

The IFC allows the use of open flames for theatrical performances, food preparation, religious ceremonies, decoration, and paint removal. Open flames are prohibited in sleeping units of Group R-2 dormitories and for cooking on combustible balconies of Group R-1 and R-2 occupancies unless they are protected by an automatic sprinkler system. Under very limited conditions, open flames are permitted in assembly (Group A) occupancies. The IFC requires an operational permit for using open flames. [Ref. 105.6.31]

When open flames are used for decorations, the fuel source cannot be liquefied petroleum gas or a liquid with a flash point temperature less than 140°F. (See Figure 3-3) If the device contains more than 8 ounces of fuel, it must be designed to be self-extinguishing and have a limited rate of fuel release if it is tipped over. The decorative flame source must be adequately secured and located so it is not an ignition source of interior finishes such as shades or curtains. [Ref. 308.3.1]

Open flames are commonly used in the table side preparation of food and beverages. These activities commonly occur in Assembly occupancies

FIGURE 3-3 An open flame decorative device

such as restaurants and night clubs—therefore, the use of open flames in an occupancy with a large occupant load requires close supervision and detailed regulations. (See Figure 3-4) The IFC limits the volume of liquid that can be dispensed to one ounce or less and limits the container volume to 1 quart. The activity must have a controlled flame height and is limited to the immediate area where the food is prepared for the consumers. Flaming foods and beverages may not be carried through the restaurant or night club. The person who prepares the flaming food or beverage is required to have a wet cloth towel to extinguish the flame in the event of an emergency. [Ref. 308.1.8]

FIGURE 3-4 Open flames used to prepare food and beverages are regulated by the IFC

VACANT PREMISES

Vacant buildings can present a significant fire threat to a community. These buildings can be used by transients for housing or for illegal activities. The building itself can be made unsafe by the theft of plumbing and electrical components manufactured from copper or other valuable materials. Thieves will open walls and shafts to remove these building materials, creating vertical paths for fire spread. To limit the risk of unwanted fires, the IFC has requirements for fire safety in vacant buildings.

Buildings that are vacated can be demolished by the jurisdiction. In many communities the jurisdiction may place a lien on the property to recover the demolition costs. Demolition generally occurs when a building is continuously used for illegal activities, is structurally unsafe, or is a fire hazard or a public nuisance. In other cases, the building may be secured or even renovated. In such cases, the securing of the building or its renovation must comply with the IBC, the *International Property Maintenance Code,* and the IFC. (See Figure 3-5) [Ref. 311.1.1]

Safeguarding a building requires that openings into the structure, such as doors and windows, are protected from unauthorized entry. (See Figure 3-6) Whenever possible, fire protection systems should be maintained in service—however, this can be difficult especially in cold weather environments which can freeze water in wet-pipe sprinkler or standpipe systems or in hot, humid environments that can cause corrosion in electronic components installed in fire alarm control units and smoke detectors. In these cases the fire code official can permit the system to be disabled, provided that combustibles and hazardous materials are removed from the building and the building's location in relation to other exposure buildings does not represent a fire hazard. In all cases, any fire-resistance construction and assemblies must be maintained in vacant buildings to limit the spread of fire. [Ref. 311.2]

The IFC authorizes the fire code official to placard unsafe buildings to warn firefighters of

FIGURE 3-5 A vacant building that is not safeguarded. From the exterior the building is unsafe and should be secured or demolished in accordance with the IFC and *International Property Maintenance Code*

FIGURE 3-6 A safeguarded vacant building

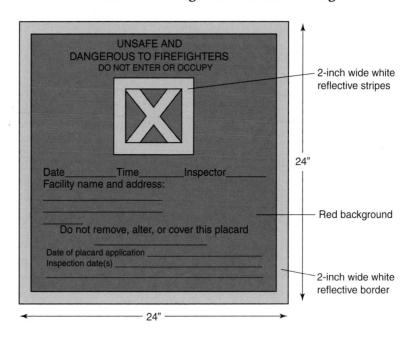

FIGURE 3-7 Firefighter building warning placard

interior hazards. (See Figure 3-7) The placard is used to indicate if a structure is safe to enter during firefighting operations or to indicate certain structural and life safety hazards to firefighters. Placards are required on all sides of a building and at entry doorways. The IFC dictates the minimum size and symbols required on the placard. [Ref. 311.5]

INDOOR DISPLAYS

Indoor displays of goods, vehicles, or exhibitions must be located and arranged so they are not an obstruction of the means of egress. The IFC prohibits the indoor display of fireworks, flammable and combustible liquids, liquefied compressed flammable gases, oxidizers, agricultural goods, and pyroxylin plastics in malls and exit access corridors or within 5 feet of any means of egress opening if the fire code official believes a fire could prevent or otherwise obstruct egress. [**Ref. 314.3**]

Vehicle displays inside of buildings must be adequately safeguarded to limit the amount of fuel and ignition sources. (See Figure 3-8) The IFC requires that such displays limit the amount of fuel to 5 gallons or ¼ of the tank volume, whichever is smaller, and that the fuel tank fill opening is sealed and the batteries are disconnected. [**Ref. 314.4**]

FIGURE 3-8 A vehicle display inside a covered mall building

Emergency Planning and Preparedness

Planning for an emergency is an important component in the overall fire and life safety of building occupants and includes prompt notification of emergency services, providing the necessary information to first responders so they can safely and quickly mitigate the hazard, clearly documenting a buildings or site's fire safety plans, and training of employees in how to respond to an emergency. Buildings or facilities storing and handling hazardous materials must provide a mechanism so the hazards of these materials are understood by individuals who work with these chemicals. Chapter 4 presents the IFC provisions that address emergency planning and preparedness, including notification of emergency forces, public assemblies and events, fire safety plans, emergency evacuation drills, and employee training and response to emergencies.

EMERGENCY FORCES NOTIFICATION

Prompt notification of emergency responders is essential to controlling an incident involving a fire or unauthorized discharge of hazardous materials. (See Figure 4-1) In the event of an emergency, the IFC requires immediate notification of the fire department and implementation of appropriate emergency plans and procedures by employees and occupants of the building. The fire code prohibits any delay in reporting a fire as well as reporting any false alarm. [Ref. 401.3]

PUBLIC ASSEMBLIES AND EVENTS

Public assemblies and events can occur inside of buildings or outdoors. Depending on the nature of the event and the number of persons estimated to be in attendance, the fire code official is granted authority to ensure that fire and life safety is maintained. One means of ensuring that safety is maintained is the use of fire watch personnel. Depending on the jurisdiction, fire watch personnel can be any approved individual, or the jurisdiction may require the use of fire department personnel. The IFC authorizes the fire code official to deploy fire watch personnel based on the anticipated number of persons or the nature of the performance, exhibition, or activity. Fire watch personnel are responsible to maintain watch for any unwanted fire and to extinguish any incipient fire, ensuring means of egress paths and openings are maintained, and to assist in evacuating people from the event. [Ref. 403.1]

Some events can have a major impact on a community, such as the arrival of dignitaries or an annual community festival. In other than assembly and educational occupancies, the IFC allows the fire code official to develop or prescribe a public safety plan for such large events to address issues such as fire apparatus access, emergency medical response, law enforcement, and similar issues. (See Figure 4-2) [Ref. 403.2]

FIGURE 4-1 Immediate notification of a fire or hazardous material release will result in a response by the fire department

FIGURE 4-2 A large community event such as this music festival may require the preparation of a public safety plan to protect festival participants

FIRE SAFETY AND EVACUATION PLANS

Certain buildings require fire safety and evacuation plans. The IFC requirements are based on the building's occupancy classification, the occupant load, or a particular building use. The plan is maintained and updated by the building owner when staff assigned to perform fire safety functions change, the building occupancy changes, or a physical change to the interior architecture occurs. The plan must be available to all employees. Fire safety plans are not required for Group S and U occupancies. [Ref. 404.2]

A fire safety and evacuation plan is required for the following occupancies and buildings:

- Group A, other than Group A occupancies used exclusively for purposes of religious worship that have an occupant load less than 2,000
- Group B buildings having an occupant load of 500 or more persons or more than 100 persons above or below the lowest level of exit discharge
- Group E
- Group F buildings having an occupant load of 500 or more persons or more than 100 persons above or below the lowest level of exit discharge
- Group H
- Group I
- Group R-1
- Group R-2 college and university buildings
- Group R-4
- High-rise buildings
- Group M buildings having an occupant load of 500 or more persons or more than 100 persons above or below the lowest level of exit discharge
- Covered malls exceeding 50,000 square feet in aggregate floor area
- Underground buildings
- Buildings with an atrium and having an occupancy in Group A, E, or M [Ref. 404.2]

The evacuation plan documents if the building will be evacuated in its entirety or partially, such as in a high-rise building where selective evacuation and relocation may be appropriate. The plan identifies key personnel who need to remain to operate critical equipment (such as smoke control or emergency power systems), procedures for determining that the partial or complete evacuation has been completed and all persons are accounted for, the primary and alternate procedures for notifying the occupants of the evacuation, and the method for notifying the fire department. If the building is equipped with an emergency voice/alarm communications system, the primary and any secondary messages that will be announced must be documented and approved. [Ref. 404.3.1]

EMERGENCY EVACUATION DRILLS

With the exception of Group H, S, and U occupancies, all other IFC regulated occupancies require one or more annual emergency evacuation drills. An emergency evacuation drill is defined as *an exercise performed to train staff and occupants and to evaluate their efficiency and effectiveness in carrying out emergency evacuation procedures*. An emergency evacuation drill is an exercise to familiarize employees or occupants with the building's fire safety and evacuation plan and is conducted to ensure the plan is properly implemented. (See Figure 4-3) Deficiencies or limitations of the plan are commonly identified during these drills, and these findings should be used to improve the fire safety and evacuation plan. [Ref. 405.2]

The frequency of emergency evacuation drills and the required participants are set forth in IFC Table 405.2 (See Table 4-1). This table establishes the number of time within a year that emergency evacuation drills are required and specifies who is required to participate. For example, IFC Section 404.2 requires a fire safety and evacuation plan for Group F occupancies with an occupant load of 500 or more persons or more than 100 persons are located above or below the lowest level of exit discharge. Under the provisions in Table 405.2, plant employees must participate in a emergency evacuation drill annually. [Ref. 405.2]

When an evacuation drill is planned, the event should be scheduled to occur at different times and under varying conditions to simulate unusual conditions that could occur. Records documenting particular aspects of the drill, such as the person who supervised the event, its date and time, the method of initiating the event, the number of persons evacuated, and the time to complete the evacuation must

FIGURE 4-3 An evacuation of a multiple story Group B office building

TABLE 4-1 Fire and evacuation drill frequency and participation (IFC Table 405.2)

Group or Occupancy	Frequency	Participation
Group A	Quarterly	Employees
Group B[c]	Annually	Employees
Group E	Monthly[a]	All occupants
Group F	Annually	Employees
Group I	Quarterly on each shift	Employees[b]
Group R-1	Quarterly on each shift	Employees
Group R-2[d]	Four annually	All occupants
Group R-4	Quarterly on each shift	Employees[b]
High-rise buildings	Annually	Employees

a. The frequency shall be allowed to be modified in accordance with Section 408.3.2.
b. Fire and evacuation drills in residential care living facilities shall include complete evacuation of the premises in accordance with Section 408.10.5. Where occupants receive rehabilitation or rehabilitation training, fire prevention and fire safety practices shall be included as part of the training.
c. Group B buildings having an occupant load of 500 or more persons or more than 100 persons above or below the lowest level of exit discharge.
d. Applicable to Group R-2 college and university buildings in accordance with Section 408.3.

maintained for review by the fire code official. In some jurisdictions the fire code official is required to be notified before such an exercise is performed. **[Ref. 405.4, 405.5 and 405.7]**

EMPLOYEE TRAINING AND RESPONSE

Employees who work in buildings or occupancies that require a fire safety and evacuation plan are required by the IFC to be trained to properly respond when a fire occurs or an evacuation is required. The minimum training must address:

- The procedures for reporting a fire or emergency,
- The life safety plan for notifying, relocating, evacuating, or sheltering employees in place,
- A review of the site plan illustrating fire department access, fire hydrant locations and the assembly point(s) for evacuees,
- A review of the floor plans including primary and secondary egress paths, areas of refuge, the location of portable fire extinguishers, manual fire alarm boxes and hoses stations,
- The identification of personnel assigned the responsibility of maintaining fire safety equipment, fire protection equipment, and the control of hazards and ignition sources,
- A review of the major fire hazards associated with the normal building use and occupancy. **[Ref. 406.1, 404.3.2]**

Training is required by the IFC as part of the new employee orientation and annually thereafter. In addition to training on implementing the fire safety and evacuation plan, some businesses will require

FIGURE 4-4 Plant employees training to correctly use a portable fire extinguisher to extinguish an incipient flammable liquid fire *(Courtesy of Tyco/Ansul Inc., Marinette, WI)*

employees to be trained in the proper use and operation of portable fire extinguishers, or businesses may maintain their own structural fire brigade. If personnel are trained to use portable fire extinguishers for incipient firefighting, the IFC requires they be properly trained and know the locations of firefighting equipment and required personal protective clothing. (See Figure 4-4) **[Ref. 406.3.4]**

Site and Building Services

Fire Service Features

Fire service features include access roadways for fire department access, a water supply for manual firefighting operations, a means of identifying the building through its address or other markings, and in certain cases, a means of accessing the building through the use of a key or access device under the exclusive control of the fire department. In buildings containing fire protection systems or a fire command center, the IFC requires the location of these system controls be identified to emergency responders.

FIRE APPARATUS ACCESS ROADS

A fire apparatus access road provides apparatus access from a fire station to a facility, building, or location. It is a general term inclusive of all other terms such as fire line, public street, private street, and access roadway. The IFC provisions specify when fire apparatus access roads are required, their design and construction, required markings, and requirement for barricades or gates that cross a fire apparatus access road. (See Figure 5-1) The IFC access road requirements are normally applied to development on private property—roads used by the public are constructed to specifications developed by the jurisdiction's Public Works or Engineering department. Design criteria that can be adopted by a jurisdiction are available in Appendix D of the 2009 IFC.

An approved fire apparatus road is required for any facility, building, or portion of a building constructed or moved into the jurisdiction. The road is located so that it extends to within 150 feet of all portions of the facility and the exterior walls of the first story of the building as measured by an approved route. The key term in applying this requirement

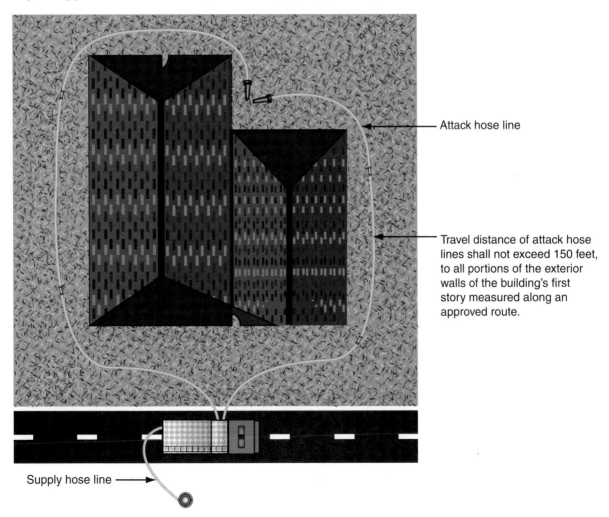

Attack hose line

Travel distance of attack hose lines shall not exceed 150 feet, to all portions of the exterior walls of the building's first story measured along an approved route.

Supply hose line

FIGURE 5-1 Measurement of the approved attack hose travel distance from a fire apparatus access road to all portions of a building

is the phrase *approved route*. An approved route is determined after considering the site topography and the geometry of the building in relation to the fire department access road. The measurement is typically commenced at the location where the fire apparatus will park on arrival and is measured along a route that can be safely used by firefighters to a point 150 feet from an engine company. The value of 150 feet is based on a typical 1½-, 1¾-, or 2-inch attack hose line carried in a horizontal cross lay, which is a conventional method of loading attack hoses on fire apparatus. The measurement stops when the attack hose can reach both sides of the building and the hose lay distance is 150 feet or less. The key to properly applying the distance is to measure the distance in the same manner the attack hose will be used by a firefighter rather than using a straight-line measurement. **[Ref. 503.1.1]**

Fire hoses are normally supplied in lengths of 50 feet for attack hoses and in 100-foot sections for water supply hoses. The IFC allows the fire code official some flexibility in the measurement. Any measurement not more than 10 to 15 feet over the 150-foot limit is acceptable in most cases, since nozzles are attached to attack hoses to deliver a fire stream that can travel 15 to 30 feet, depending on the nozzle setting and the discharge pressure.

The IFC allows the distance from a fire apparatus access road to a building to be increased when it is protected with an approved automatic sprinkler system that complies with NFPA 13, *Installation of Sprinkler Systems,* NFPA 13R, *Sprinkler Systems for Residential Occupancies up to and Including Four Stories in Height,* or NFPA 13D, *Sprinkler Systems for One- and Two-Family Dwellings and Manufactured Housing.* The amount that the distance can be increased is not specified in the IFC—the decision rests with the fire code official. In instances where fire apparatus access roads cannot be constructed because of topography, waterways, or non-negotiable grades, the fire code official can allow the use of an alternative fire protection design in lieu of required access roads. Code officials commonly accept a building protected throughout by a NFPA 13, 13R, or 13D automatic sprinkler system. Access roads are not required where there are no more than two Group R-3 or U occupancies. **[Ref. 503.1.1, exception]**

Fire apparatus access roads must be constructed to the requirements of the jurisdiction and the IFC. The IFC requires fire apparatus access roads have a minimum unobstructed width of 20 feet and an unobstructed clearance of not less than 13 feet, 6 inches. Road surfaces must be designed to support the imposed load of the fire apparatus and constructed of materials that are resistant to any weather conditions. The maximum permitted grade of the roadway as well as the inside and outside road turning radii is required to be based on the fire department's apparatus. The IFC limits the dead-end roads to a maximum length of 150 feet before a turnaround is required. (See Figure 5-2) **[Ref. 503.2]**

The fire code prohibits obstructing the road with vehicles or stationary objects such as dumpsters. Fire apparatus access roads require markings using approved signs that demarcate its purpose and prohibit any obstructions. (See Figure 5-3) The design and placement of signs is

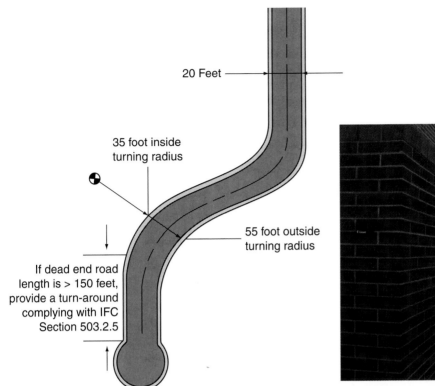

20 Feet

35 foot inside turning radius

55 foot outside turning radius

If dead end road length is > 150 feet, provide a turn-around complying with IFC Section 503.2.5

FIGURE 5-2 The IFC limits the length of any dead end road to a maximum distance of 150 feet

FIGURE 5-3 Fire lanes must be identified to demarcate that its primary function is for fire apparatus access *(Courtesy of the City of Phoenix (AZ) Fire Department)*

generally governed by a state's traffic code, which provides for consistency in the sign's appearance and is also legally admissible in court if the jurisdiction issues citations for traffic violations. **[Ref. 503.3, 503.4]**

Security gates across fire department access roadways can slow the emergency response. An approved gate installation requires an alternate means to ensure its operation during an emergency, such as a source of standby power or a manual releasing mechanism. Gates across access roads require the approval of the fire code official. **[Ref. 503.5]**

The IFC requires that all gates comply with the requirements of American Society of Testing and Materials standard F 2200-05, *Standard Specification for Automated Vehicular Gate Construction*, ASTM F 2200 establishes general requirements for all automated vehicle gates and construction requirements for five basic designs commonly found in the United States: horizontal slide, horizontal swing, vertical lift, vertical pivot, and overhead pivot. The ASTM standard also divides gates into different applications based on the functions of the building the gate is serving:

- Class I—One- and two-family dwellings, up to four dwellings.
- Class II—Commercial uses where general public access is desired. An example is an apartment community or a subdivision of one- and two-family dwellings.

- Class III—Industrial uses not intended to serve the public. An example is a manufacturing plant where public access is limited.
- Class IV—High-security locations, including those with guard service or security surveillance.

The class of gate installed across a fire department access road is not specified by the IFC. The gate classification is dictated by the installing contractor and is selected to ensure that its design, construction, and installation comply with the requirements in ASTM Standard F 2200 and the IFC.

ASTM F 2200 was developed to ensure that persons near an automatic gate are not exposed to a potential entrapment or entanglement hazard. This includes the location of the automatic gate in relation to a fixed object such as a pipe bollard or building—in this case, the operation of a gate could create a potential crush injury to a person between the gate and the fixed object. (See Figure 5-4) Pedestrian gates are outside of the scope of ASTM F 2200, and automated vehicular gates should not be used as pedestrian gates. [Ref. 503.6]

ACCESS TO BUILDINGS

Ensuring that firefighters can quickly access the building completes one step in the response. To complete the response, access must be gained to the building. The IFC requires that exterior doors or openings required by the IBC be maintained accessible for use by emergency responders. [Ref. 504.1]

The owner or occupants will have concerns over the safety and security of the individuals who use and occupy the building and its contents. Sometimes the level of security can conflict with the needs of emergency responders to access the structure and its occupants.

FIGURE 5-4 Automatic gate installed across a fire department access road *(Courtesy of the City of Phoenix (AZ) Fire Department)*

One method that is generally regarded as a reasonable means of ensuring that the building's security is maintained while allowing for rapid firefighter access is the installation of a fire department key box. (See Figure 5-5) The key box can only be opened by a master key that is carried by the company officer or can only be accessed when permission is electronically granted by the fire department's communication center. Many of the key box manufacturers have their equipment evaluated by a nationally recognized testing laboratory to demonstrate that it is resistant to burglars. The location of the key box, the required number of keys, and the manufacturer of the key box must be approved by the fire code official. **[Ref. 506.1]**

The IBC requires that one of the building stairways has a means of accessing the roof when the building height is four or more stories above the grade plane. Roof access is not required when the roof is pitched and the slope is greater than 4 units vertical in 12 units horizontal, which equals a 33 degree slope. Roof access is provided because it can be used as a location to deploy fire streams to protect the structure from an exposure building fire. When access is provided, it can be through a penthouse or through a roof hatch. The stairway must be marked to indicate that it has roof access. (See Figure 5-6) **[Ref. 504.3]**

HAZARDS TO FIREFIGHTERS

While manual firefighting is an inherently hazardous activity, it is safely performed daily because of the required certifications and training, assessment and implementation of fireground risk management, personal protective equipment, and the incident command system used to manage the emergency. Even with these controls and procedures, certain building features such as open building shafts may be present.

FIGURE 5-5 Fire department key box

FIGURE 5-6 The doorway is identified to indicate it serves a stairway with roof access. Note that the exterior doorway is accessible by using the fire department key box

FIGURE 5-7 A shaft way that is accessible from the exterior is required to be marked to indicate its location to firefighters

In some cases where security concerns are significant, a building may be equipped with a device intended to incapacitate or injure a burglar. However, in the event of a fire, such a device could accidentally cause the injury or death of firefighters. The IFC has specific requirements for protecting firefighters from certain hazards that may be found in buildings.

Building shaftways are provided for elevators, material handling equipment, or for utilities such as electrical conductors or air ducts. Shaft ways present a fall hazard to firefighters because they generally are constructed to access every floor of a building. An accidental fall in a shaft way can easily injure or kill a firefighter. To prevent the potential from an accidental fall in a shaft way, the IFC requires warning signs to identify shaft openings. If the shaft is accessible from the building exterior, a sign is required at a location that indicates the shaft's location. (See Figure 5-7) **[Ref. 316.2.1]**

FIRE PROTECTION WATER SUPPLIES

Any facility or building constructed in a jurisdiction that has adopted the IFC requires a fire protection system water supply capable of delivering the required fire flow for manual firefighting operations. The source of the water supply can be a public water distribution system, an underground well supplied from a fire pump, a water storage tank, a reservoir, or private fire service mains connected to a public water system. **[Ref. 507.1]**

Fire flow is defined in IFC Appendix B as *the flow rate of a water supply, measured at 20 pounds per square inch (psi) residual pressure that is available for firefighting.* The acceptable method used to determine fire flow rests with the fire code official. Many jurisdictions chose to adopt IFC Appendix B because it specifies the required fire flow based on the building's height, area, and construction type. The jurisdiction may use the fire flow methods developed by the National Fire Academy or Iowa State University. In rural jurisdictions where a conventional water supply system is unavailable, Appendix B permits the fire code official to use mobile tankers as the mechanism for water supply delivery provided the tankers and the delivery mechanism complies with the *International Wildland-Urban Interface Code* or NFPA 1142, *Standard on Water Supplies for Suburban and Rural Firefighting.* **[Ref. 507.3, B103.3]**

When a municipal or private water supply system is the source for fire flow, a flow test is performed to demonstrate that the system is capable satisfying the water supply. A flow test uses a test fire hydrant and a flow fire hydrant to measure the static pressure and residual pressure of the water supply system as well as the available flow rate, expressed in gallons/minute (GPM). (See Figure 5-8) Static pressure is the available pressure of the water supply system with the water at rest. The static pressure can be developed by water pumps or the pressure of water in elevated tanks. Residual pressure is the available pressure when water is discharged from the flowing fire hydrant. The residual pressure is measured at the same time water is flowing from the flow hydrant. These three values are used to calculate the available fire flow. The IFC requires the fire flow be calculated at a residual pressure of 20 PSIG. 20 PSIG is the lowest residual pressure allowed by many state health departments and water regulatory authorities. Pressures below this value introduce the possibility of a water main collapse or worse, backflow of untreated sewage or wastewater into the potable water supply. [**Ref. 507.4**]

Fire protection water supplies are commonly supplied from private hydrants and private fire service mains are constructed on private property to supply the required fire flow. The IFC requires the construction of private fire protection water mains be in accordance with NFPA 24, *Standard for Fire Service Water Mains*. A building must be located within 400 feet of a fire hydrant on a fire apparatus access road. The distance is measured along an approved route using the same method prescribed by the fire code for locating buildings in relation to fire apparatus access roads. The distance is measured using a path that fire apparatus will unload water supply hoses on the roadway instead of a straight line measurement from the fire hydrant to the building. When a building is protected by an automatic sprinkler system installed in accordance with NFPA 13 or NFPA 13R, the fire code official can increase the distance requirement to 600 feet. For Group R-3 and U occupancies, the travel distance is also 600 feet. [**Ref. 507.5.1**]

A fire protection water supply system requires annual maintenance to ensure the system will deliver the required fire flow. (See Figure 5-9) The IFC references the requirements in NFPA 25, *Standard for the Maintenance of Water-Based Fire Protection Systems* and specifies the following frequencies for inspection, testing, and maintenance:

- Private fire hydrants: annual flow test and maintenance
- Private fire protection water mains: a flow test every 5 years
- Strainers installed in private fire protection water mains: inspection and maintenance after each use. [**Ref. 507.5.3**]

Code Basics

The static pressure, residual pressure, and flow rate (measured in gallons/minute) are required to determine the fire flow. Fire flow is calculated at a residual pressure of 20 PSIG. ●

FIGURE 5-8 Flow testing of a fire hydrant to determine its available water supply

FIGURE 5-9 Flushing of a private fire hydrant to remove and discharge any sediment or debris in the private fire protection water main

To ensure that fire hydrants are always accessible, the IFC prohibits their obstruction and requires a minimum 3-foot clearance around their circumference. When a fire hydrant is located in an area subject to impact by motor vehicles, a means of vehicle impact protection complying with the IFC is required. (See Figure 5-10) **[Ref. 507.5.5 and 507.5.6]**

EMERGENCY RESPONDER RADIO COVERAGE

Public safety radio systems in numerous regions of the United States have been converted from or are in the process of converting from analog signaling systems to digital signaling systems. Because they are digital systems, the portable radios used by firefighters and emergency responders require less power for transmitting audio or electronic signals compared with that required for the former analog systems. (See Figure 5-11) These systems also offer greater functionality, because each radio is assigned its own digital address within the public safety radio system, based on the user's roles and responsibilities at emergency incidents. These systems provide for interoperability, meaning a firefighter can now directly communicate with police officers, emergency medical personnel, and any other person assigned to a particular emergency incident. This interoperability can be extended to other emergency responders in one or more regions of a state to one or more states or areas in the United States.

The IFC emergency responder radio coverage provisions are concerned with the reliability of portable radios used inside buildings. Portable radios are used to communicate between other emergency responders, the incident commander, or the public safety communications center. Building construction features and materials can absorb

FIGURE 5-10 A fire hydrant with vehicle impact protection

FIGURE 5-11 A firefighter using a portable radio inside a building

or block the radio frequency energy used to carry the signals inside or outside of the building. Blockage or absorption of the radio frequency signal can prevent a critical message from an emergency responder from being received and acknowledged. Depending on the incident, this loss of information can place other emergency responders in greater danger or may prevent an injured or disoriented emergency responder from communicating for assistance.

Section 510 and Appendix J address requirements for in-building coverage of emergency responder radios. The requirements in this section apply to analog and digital radio systems. Section 510 contains requirements for all buildings. Section 510.1 requires that all buildings have approved radio coverage for emergency responders within the building. (See Figure 5-12) Approved radio coverage is based on the ability of the public safety communication system to transmit a signal inside and outside of the building. Section 510.2 requires that a minimum signal strength of −95 dBm be received by radios inside the building and a minimum signal strength of −100 dBm be transmitted outside the building. The radio coverage is considered acceptable if these signal strengths can be maintained in 95% of all areas on each floor of a building. [Ref. 510.1 and 510.2]

If testing finds that these signal strengths are not satisfactory, the IFC authorizes the code official to require the installation of a wired communication system. When this exception is applied, the concurrent approval of the fire and building code officials is required. The IFC does not offer any criteria as to what constitutes an acceptable wired communication system. Examples of an acceptable system may be a fire department communications system complying with Section 907.6.2.2 or a building's telephone system. [Ref. 510.1]

Exception 2 of Section 510.1 allows the fire code official to waive the requirement when it is determined that emergency responder radio coverage is not needed. Exception 2 does not give any criteria as to when

FIGURE 5-12 The performance criteria in Section 510.1 ensure that signals transmitted from the public safety communication center are received by emergency responders operating inside a building

these requirements are applicable or which buildings can be exempted. Discussions with public safety radio professionals found that based on current radio technologies, these requirements should be applied in any building with one or more basement or below-grade building levels, any underground building, or any building more than five stories in height. In most wood frame or mixed construction Group R-1 and R-2 occupancies, single-family dwellings, townhouses, and buildings with an area less than 50,000 square feet without basements, there is little concern for loss of radio signal strength inside the building or inability to transmit to an outdoor receiver.

The IFC also specifies the requirements for emergency responder radio coverage in existing buildings. As previously noted, when radio coverage is not adequate, Section 510.1 Exception 1 permits the installation of a wired communication system. However, if the wired communication system cannot be repaired or is undergoing replacement, Section 510.3 allows the code official to require the installation of a building radio amplification system. If an existing building is known to have inadequate radio coverage based on the requirement in Section 510.2, Exception 2 of Section 510.3 allows the code official to establish a time frame for compliance with Section 510.1. This could be either a wired communication system or a building radio amplification system. Prior to construction, there are times when it will be difficult to ensure that radio coverage will not be affected by the building itself. This is the reason this requirement is also applicable to existing facilities. [Ref. 510.3]

Requirements for an in-building amplification system are found in Appendix J, *Emergency Responder Radio Coverage*. Appendix J is not mandatory unless the jurisdiction adopts it by reference. Code officials should consider adopting Appendix J unless the jurisdiction has its own criteria for the in-building performance of its analog or digital radio system.

Determining whether a building has adequate or inadequate radio coverage is most easily accomplished by having personnel perform functional radio tests inside areas of the existing building. For buildings under construction, design professionals may issue a letter of intent of compliance to the code official that documents that the owner will comply with the requirements and install either a wired communication system or a signal booster system if the functional radio test fails to transmit a signal outside the building or to other radios inside the building after the construction is substantially completed. Some jurisdictions are requiring a third-party review of the building features before issuing a building permit, but this can be difficult because the radio frequency signal absorption and deflection characteristics of different building materials are not completely understood by the engineering community. If code officials wish to perform a functional test, Section J103.3.5 authorizes the code official to enter building to perform the tests at any reasonable time. [Ref. 510.4]

In new buildings, the fire code official can apply the IFC acceptance testing criteria as a means of confirming that portable radios will provide adequate emergency responder radio coverage. Persons performing these

You Should Know

A building constructed in a jurisdiction that has adopted the IFC will require fire apparatus access roads, a water supply capable of supplying the required fire flow, and a means of access into the building. Private fire hydrants and fire protection water mains may be required as a source of fire flow. All buildings require emergency responder radio coverage. ●

tests should have a valid Federal Communication Commission General Radio Operators License and a certification as to their understanding of the in-building amplification equipment as specified in Section J103.2.3. The individual performing the tests should be using a radio frequency spectrum analyzer and its associated equipment. Code officials may not be completely confident in witnessing such tests and should consider applying the requirements in Section 106.2 to treat this test as a special inspection. [**Ref. 510.2**]

Building Systems

Building systems include fuel-fired appliances, standby and emergency power, and control of elevators by emergency responders. The provisions in IFC Chapter 6 set forth minimum requirements for these building systems. Some of the systems use hazardous materials that are exempt from regulations in IFC Chapter 27. (See Part 6 of this book.) These systems are not exempted—rather, they are required to comply with the IFC requirements, its adopted standards, and the requirements in the IMC and IBC.

FUEL-FIRED APPLIANCES

With the exception of internal combustion engines and portable appliances such as ice melting or weed-burning torches, the IFC regulates the installation and operation of fixed fuel gas appliances. An appliance is an apparatus or device using fuel gas or fuel oil to produce light, heat, power, refrigeration, or air conditioning. The IFC requires that all appliances be installed in accordance with the requirements of the *International Fuel Gas Code* (IFGC) as well as the requirements of the IMC. **[Ref. 603.1, 603.1.2]**

In the U.S., the two fuel gases commonly consumed are natural gas and liquefied petroleum gas (LP-Gas). Natural gas is a mixture of flammable gases whose primary constituent is methane. LP-Gas is a mixture of ethane, methane, propane, and butane. Natural gas and LP-Gas are colorless and odorless flammable gases. The density of natural gas makes it lighter-than-air when it is released, while LP-Gas is heavier-than-air. The U.S. Department of Transportation requires both gases be odorized when they are shipped using a pipeline. An odorant like methyl mercaptan is added so any leak is detectable by a distinctive odor when the volume of the release is at 25% of its lower flammable limit.

Fuel oil is a hydrocarbon distillate with a flash point temperature of more than 100°F, which classifies the material as a combustible liquid based on the classification criteria in IFC Chapter 34. (See Chapter 18 for an explanation about classification of flammable and combustible liquids.) Fuel oils include kerosene and diesel fuel.

All fuel-fired equipment is required to be installed in accordance with the manufacturer's instructions and modification of the equipment must be performed in accordance with the requirements of the original equipment manufacturer. Adequate access is required around equipment so it can be maintained. (See Figure 6-1) The IFGC and IMC both contain extensive requirements for equipment access. **[Ref. 603.1.2 and 603.1.5]**

Code Basics

Liquefied petroleum gas (LP-Gas) and methane (natural gas) are two forms of fuel gas. All fuel gases are odorized so leaks can be easily detected. Kerosene and no. 2 diesel fuel are examples of fuel oils. The IMC requires fuel oils have a flash point temperature of more than 100°F. ●

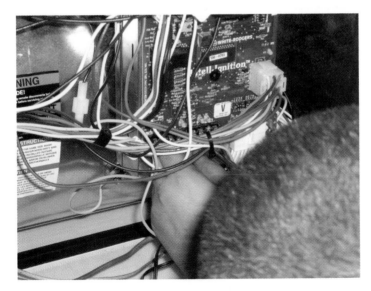

FIGURE 6-1 Fuel-fired equipment must be installed with adequate access to facilitate maintenance

FIGURE 6-2 Waste oil burners
(Courtesy of Clean Burn Energy Systems, Leola PA)

Fuel oil is used as a source of heat and to operate engine driven fire pumps or generators. Petroleum products used in fuel-fired equipment must meet the original equipment manufacturer's specifications, and the use of a petroleum product that is contaminated with gasoline is prohibited by the IFC. In Group F, M, and S occupancies, the business may choose to use heaters that consume waste oil. Waste oil heaters and boilers are commonly found in repair garages and manufacturing plants using petroleum formulated cutting fluids. (See Figure 6-2) Waste oil fuel-fired appliances are required by the IFC to be listed as heat recovery devices. These heaters are limited to the use of used crankcase oil up to Society of Automotive Engineers 50 weight, used transmission and hydraulic fluids, and number 2, 3, and 4 fuel oils. If waste oil heaters are installed in a repair garage, they also must comply with the IFC Chapter 22 and NFPA 30A, *Automotive and Marine Service Station Code* requirements. **[Ref. 603.1.4]**

The amount of fuel oil allowed inside of a building is dependent on the type of storage tank and the level of fire protection. When stored aboveground, the IFC permits the installation of the storage tank outside or inside a building. Fuel oil storage outside of buildings is limited to a volume of 660 gallons. If more than 660 gallons of fuel oil is required, the design of the storage tank and fuel oil delivery system must comply with the requirements of NFPA 31, *Standard for the Installation of Oil-Burning Equipment*. NFPA 31 defers to the requirements of NFPA 30, *Flammable and Combustible Liquids Code* for tanks storing more than 660 gallons of fuel oil. **[Ref. 603.3.1]**

Requirements for the indoor storage of fuel oil are based on the storage amount. Storage of less than 660 gallons of fuel oil is allowed in one or more tanks. If the quantity exceeds 660 gallons, the fuel oil storage is allowed in any underground or aboveground tank permitted by IFC Chapter 34. Up to 3,000 gallons is permitted aboveground and inside a building when the fuel oil is stored inside of protected aboveground storage tank (PAST). (See Figure 6-3) **[Ref. 603.3.1]**

FIGURE 6-3 A 1,000 gallon protected aboveground storage tank supplying an engine-driven generator

A PAST is a shop-fabricated aboveground storage tank that has been subjected to a fire test that replicates exposure to a two-hour flammable liquid pool fire. PASTs are constructed with integral secondary containment and are evaluated for vehicle impact and bullet resistance. All openings on a PAST are located at the top of the storage tank, which reduces the potential for liquid leaks—openings below the liquid level in these tanks are prohibited by the IFC. These tanks are listed as meeting the requirements of UL Standard 2085, *Protected Aboveground Tanks for Flammable and Combustible Liquids.* [Ref. 603.3.2.1]

To limit the potential of a fire involving the PAST, the floor level housing the storage tank must be protected by an approved automatic sprinkler system that complies with the requirements in Section 903.3.1.1. The function of the PAST is limited to the supply of fuel oil to fuel-fired appliances, including generators—it cannot be used for any other purpose. The PAST cannot be located more than two stories below the building's grade plane. Fuel oil piping must be liquid tight and comply with the requirements of the IMC. (See Figure 6-4) [Ref. 603.3.2.1, 603.3.2.5]

Fuel-fired appliances are designed to operate as either vented or unvented appliances. They are commonly used to heat a room or an area of a room. A vented appliance directly vent combustion gases produced by the fuel gas or fuel oil outside of the building. These devices use air in the room they are located in as its source of supply air.

Unvented heaters also use indoor air as their source of combustion air. However, these heaters vent the combustion byproducts into the room. Most portable unvented heaters use kerosene or LP-Gas as fuel. (See Figure 6-5) When these heaters are properly operated and

> ## Code Basics
>
> The amount of fuel oil allowed inside a building is limited by the IFC based on the type of storage tank. Aboveground and inside of buildings, not more than 3,000 gallons of fuel oil can be stored. ●

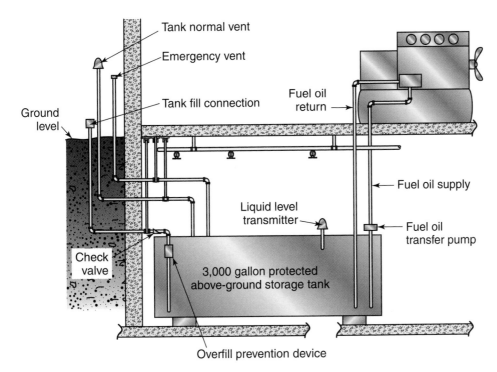

FIGURE 6-4 A 3,000 gallon PAST inside a building

FIGURE 6-5 A portable unvented appliance used inside of a work area of a Group B occupancy

maintained, these systems can be up to 98% efficient. Although even with this high efficiency, the constant use of unvented heaters inside a building can introduce carbon monoxide into a room if fresh makeup air is not introduced. Carbon monoxide molecules have a higher affinity to attach to red blood cells than oxygen. If an atmosphere contains enough carbon monoxide in the air, the carbon monoxide can displace the oxygen, preventing it from being carried into the body. Carbon monoxide poisoning can cause injury or death.

Because of the concerns over carbon monoxide poisoning, as well as issues of fire safety since these heaters emit radiant and convective energy, the IFC has specific provisions concerning portable unvented heaters. Unvented heaters are prohibited in Assembly, Educational, Institutional, and Residential occupancies. Listed and approved heaters are allowed in one- and two-family dwellings. [Ref. 603.4]

Portable unvented heaters are not permitted to be located in or to obtain combustion air from sleeping rooms, bathrooms, or closets. [Ref. 603.4.1]

Another form of unvented heater are portable outdoor gas-fired appliances. These heaters are especially popular at outdoor seating areas of restaurants and similar occupancies. Being unvented appliances, they are not designed or listed for indoor use. The IFC prohibits the use of portable outdoor gas-fired heating appliances inside buildings, tents, canopies, and membrane structures. (See Figure 6-6) Use of these heaters is also prohibited on exterior balconies of apartment buildings and similar multifamily residential areas, as prescribed in NFPA 58, *Liquefied Petroleum Gas Code*. NFPA 58 prohibits the use or storage of liquefied petroleum–gas (LP-Gas) containers on exterior balconies of apartments when the container volume is greater than 1.08 pounds of propane. [Ref. 603.4.2.1.1]

Portable outdoor gas-fired heating appliances are required be listed and approved by the fire code official. (See Figure 6-7) For patio heaters, the applicable standard is American National Standards Institute (ANSI)

FIGURE 6-6 A portable outdoor gas-fired heating appliance located inside a building. The installation is a violation of Section 603.4.2.1.1

Z83.26, *Standard for Gas-Fired Outdoor Infrared Patio Heaters.* One safety feature required by the ANSI standard is the connection between the LP-Gas cylinder and the hose supplying the appliance's burner. The standard requires that the hose connected to the appliance be equipped with a Compressed Gas Association (CGA) 790 fitting. A CGA 790 fitting provides three separate safety features. First, the fitting has a thermal link that is designed to activate at temperatures of 200° to 250°F that stops the flow of LP-Gas in the event of a fire. Second, the fitting requires a positive connection to the cylinder before LP-Gas can flow into the appliance. Third, the fitting is equipped with an internal excess flow control valve. An excess flow control valve is designed to stop the flow of a gas or liquid in the event of hose or pipe rupture. [Ref. 603.4.2.2.1]

A minimum five-foot separation is required between the appliance and buildings, combustible awnings or overhangs, decorations, and exits and exit discharges. [Ref. 603.4.2.1.2, 603.4.2.1.3 and 603.4.2.1.4]

FIGURE 6-7 A portable outdoor gas-fired appliance *(Courtesy of Infrared Dynamics, Yorba Linda, CA)*

EMERGENCY AND STANDBY POWER SYSTEMS

Emergency and standby power systems are required by the IFC and IBC to provide reliable, second source of electric power to buildings, uses, or facilities. These systems are required in buildings that represent a significant life safety challenge, can present complications to emergency responders, or contain certain classes of hazardous materials. The IFC provisions for these systems specify when they are required, system installation and performance requirements based on the building's use, and the inspection, testing, and maintenance of the power source and its transfer switch. The requirements for design and installation of emergency and standby power systems are specified in NFPA 70, *National Electrical Code©* (NEC©), NFPA 110, *Emergency and Standby Power Systems,* and NFPA 111, *Stored Electrical Energy Emergency and Standby Power Systems.* NFPA 110 addresses backup power systems that connect to an alternative energy grid or to engine-driven generator sets, and NFPA 111 addresses stored energy sources such as stationary storage batteries. [Ref. 604.1]

Emergency power and standby power are defined terms in the NEC© that explain the purpose and functions of these systems. The NEC© has different requirements for emergency and standby power systems—the standard does not designate the occupancies or uses where these auxiliary power systems are required. The IFC designates the occupancy or use and type of auxiliary power system that is required—in cases where the code does not specifically indicate if standby or emergency power is required, the determination of the acceptable form of auxiliary power rests with the design professional and the fire code official. Table 6-1 summarizes the occupancies, buildings, and uses that require emergency power, standby power, or both. (See Figure 6-8) [Ref. 604.2]

Emergency power systems are specified when the interruption of electrical current would produce very serious life safety or health hazards. Means of egress exit sign lighting and exit illumination, electric

Code Basics

Fuel-fired appliances can be vented or unvented. The IFC prohibits the installation of unvented heaters in most occupancies because of the potential for carbon monoxide poisoning. ●

TABLE 6-1 Occupancies, buildings, and uses requiring emergency and standby power

Where required	Type of required system	IFC section number
Group A occupancies with an occupant load of 1,000 or more	Emergency	604.2.1
Smoke control systems	Standby	604.2.2
Exit signs	Emergency	604.2.3
Means of egress illumination	Emergency	604.2.4
Accessible means of egress elevators	Standby	604.2.5
Accessible means of egress platform lifts	Standby	604.2.6
Horizontal sliding doors	Standby	604.2.7
Semiconductor fabrication facilities	Emergency	604.2.8
Membrane structures—exit signs	Emergency	604.2.9
Permanent membrane structures—auxiliary inflation systems	Standby	604.2.9
Hazardous materials	Emergency or Standby	604.2.10
Highly toxic and toxic materials	Emergency	604.2.11
Organic peroxides	Standby	604.2.12
Covered mall buildings	Standby	604.2.13
High-rise buildings	Emergency or Standby, depending on the electrical load's function	604.2.14
Underground buildings	Emergency or Standby, depending on the electrical load's function	604.2.15
Group I-3 occupancies	Emergency	604.2.16
Aircraft traffic control towers	Standby	604.2.17
Elevators	Standby	604.2.18

FIGURE 6-8 This Group A indoor sports auditorium requires emergency power because its occupant load is more than 1,000

fire pumps, and elevator car illumination in high-rise buildings are electrical loads that must be connected to emergency power circuits. Electrical power feeder-circuit equipment and wiring for emergency power systems required protection using a passive form of fire protection such as listed thermal barriers for electrical components or protecting the area housing the equipment and wiring with an automatic fire-extinguishing system. The wiring generally must be kept independent of other building wiring. The NEC© prohibits the connection of non-emergency power loads to the emergency power circuits.

Requirements for standby power systems are not as rigorous when compared to the NEC© emergency power requirements. The IFC requires standby power systems when the interruption of normal electrical service could create hazards or interrupt firefighting or rescue operations. (See Figure 6-9) Electrical loads required to be connected to standby power branch circuits include smoke control systems, elevators, and platform lifts used as components of an accessible means of egress, and refrigeration equipment to control the storage temperatures of certain organic peroxides. Fire protection of the equipment providing standby power is not required by the NEC©.

FIGURE 6-9 The IFC requires a platform lift used as part of an accessible means of egress be connected to an approved standby power source

Emergency and standby power systems are a system of wiring, equipment controls, and an energy source that provide a safe and reliable source of electric power. (See Figure 6-10) The energy source can be stored batteries or it may be an indoor or outdoor generator set. The other critical component in an emergency or standby power system is the automatic transfer switch (ATS). The ATS is a listed electric switch that transfers the connected emergency or standby power circuits to the auxiliary energy source in the event normal electric power is disconnected. It is required to be listed for either emergency- or standby-power service and must be designed to safely carry the entire electrical load of all the circuits to which it is connected. An ATS must be designed so it can be manually switched between utility power and auxiliary power in case the automatic function of the switch fails when a power source transfer is required. The NEC© requires an ATS used for emergency power service operate within 10 seconds after power is disconnected and within 60 seconds when the switch is serving a standby power source, which is consistent with the IFC requirements for ATS devices in high-rise and underground buildings. **[Ref. 604.2.14.3 and 604.2.15.2]**

Failures of emergency and standby power systems generally result from a lack of testing and exercising of the ATS or of the energy source, such as scheduled load testing of an engine-driven generator. The IFC requires that an approved maintenance and testing schedule be prepared and implemented to ensure that these systems are available for service in the event that utility power is lost. Such a schedule must comply with the inspection and testing requirements in the NEC©, NFPA 110, and NFPA 111. The results of inspections and tests must be documented

Code Basics

Emergency power systems are required when the loss of electrical service could result in serious life safety or health hazards. Standby power systems are required when the loss of electric power could create hazards to or interrupt firefighting or rescue operations. ●

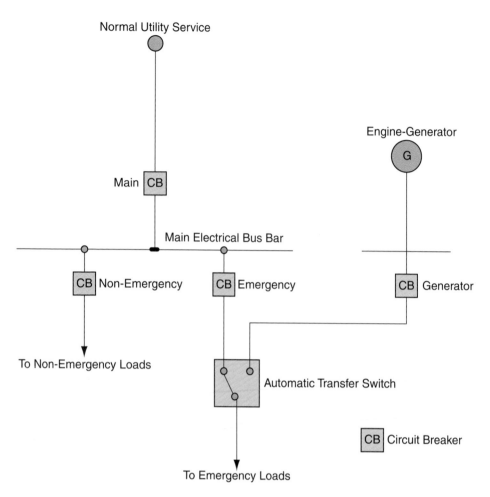

FIGURE 6-10 One-line diagram of a engine-driven generator and automatic transfer switch used as a emergency power source

FIGURE 6-11 Exercising the automatic transfer switch and load testing of an engine-driven generator are important steps in maintaining emergency and standby power systems

and be maintained available for review by the fire code official. The maintenance and inspection program must include the ATS. **[Ref. 604.3.3]**

The fire code requires that the power source, such as an engine-driven generator or a battery group, be tested under load in accordance with NFPA 110 or NFPA 111. Load testing is beneficial to engine driven generators because during the first 12 to 24 months, full or partial loading of the engine help to set piston rings, which is important to the overall performance of the generator under a full electrical load. (See Figure 6-11) If the engine piston rings are not completely set and the generator is subjected to a full design load, the engine performance can decay, causing loss of power to one or more critical loads. The tests prescribed in the IFC require tests of the primary auxiliary power source and its ATS. As an option, the standby or emergency power system can be used for load shaving. Load shaving is a practice where utility power is disconnected to a part or all of a building and a generator is used to supply electricity. Load shaving is considered equal to load testing and can be used to demonstrate compliance with the maintenance plan. [**Ref. 604.4, Exception 1**]

ELEVATOR RECALL AND MAINTENANCE

The only buildings that require an elevator are those which are five or more stories in height to maintain an accessible means of egress for mobility-impaired persons. When a building is more than 120 feet above the lowest level of fire department access, the IBC requires a fire service access elevator. In all other buildings, elevators are provided to facilitate the movement of the building occupants, and, as such, they must be installed in accordance with the applicable IBC requirements. [**Ref. 1007.2.1**]

Requirements for the design, construction, and testing of elevators are contained in ASME/ANSI A17.1, *Safety Code for Elevators and Escalators*. ANSI A17.1 requires all new elevators to be equipped with Phase I and Phase II firefighter service features. In Phase I service, a smoke detector is installed in each elevator lobby. (See Figure 6-12) Activation of the smoke detector causes the elevator to recall to the designated floor, which is usually the ground floor of a building. Upon arrival, the cab door opens and the elevator is no longer operable by the occupants. In any building that is equipped with an elevator having a travel distance of more than 25 feet, the IFC requires that it be equipped with Phase I firefighter service. This requirement is applicable to all new and existing buildings. In the event a smoke detector fails or if emergency responders wish to use the elevator such as for the transportation of equipment in the treatment of a patient, a key switch is provided in the elevator lobby. Activation of the key switch captures the elevator and recalls it to the floor level where the switch was activated. [**Ref. 607.1**]

FIGURE 6-12 A Phase I firefighter service feature

FIGURE 6-13 Phase II firefighter service feature

FIGURE 6-14 A pictorial sign is required at each elevator lobby instructing occupants to use the exit enclosure stairs during a building fire

Phase II firefighter service allows firefighters to control the elevator and travel to any floor in the elevator bank. (See Figure 6-13) The operating controls are located inside of the elevator car. When the elevator is placed into the fire service mode, the elevator can only be operated by personnel in the elevator car. To return the elevator to normal service, the elevator must be reset using the Phase I switch. **[Ref. 607.1]**

In the event that a lobby smoke detector activates, all of the elevators in the bank will be recalled to the designated floor. Elevators are generally not permitted to be used as a means of egress component by the IBC. To ensure that occupants do not attempt to use an elevator as a means of egress, a standard pictorial sign is required in each elevator lobby that instructs the occupants to use exit stairs when a building is being evacuated. (See Figure 6-14) **[Ref. 607.2]**

COMMERCIAL KITCHEN HOODS

Commercial cooking appliances are used in a commercial food service establishment for cooking food. Commercial cooking appliances require a local exhaust ventilation system to remove grease vapors, steam, smoke, or odors produced during food preparation, cooking or cleaning activities. The requirements for exhaust systems are set forth in the IMC, and commercial cooking appliances require a Type I hood. A Type I hood is designed for the removal of grease-laden vapors and smoke. (See Figure 6-15) **[Ref. 609.1]**

Type I hoods and their exhaust ducts will collect the fats produced during the cooking processes. The amount of grease and smoke that is generated is dependent on the method of cooking, the type of protein being cooked and the intensity of the cooking operations. Accordingly, the IFC establishes prescriptive inspection frequencies in Table 609.3.3.1 (See Table 6-2).

FIGURE 6-15 A Type I hood in a cooking area using solid fuel and fuel gas

TABLE 6-2 Commercial cooking system inspection frequency (IFC Table 609.3.3.1)

Type of cooking operation	Frequency of inspection
High-volume cooking operations such as 24-hour cooking, charbroiling, or wok cooking	3 months
Low-volume cooking operations such as places of religious worship, seasonal businesses, and senior centers	12 months
Cooking operations utilizing solid-fuel-burning cooking appliances	1 month
All other cooking operations	6 months

The most restrictive inspection frequency is for cooking operations using solid fuels, such as barbeque pits and meat smokers. When appliances use a solid fuel such as wood or charcoal, a minimum monthly inspection frequency is required. Cooking operations using charbroiling or woks require a minimum 3-month inspection frequency, as do high-volume cooking operations found in 24-hour restaurants. The frequency of inspection is reduced to 12 months for cooking activities involving seasonal businesses, places of worship, and facilities that provide care for the elderly. All other cooking operations are subject to a 6-month inspection frequency. Depending on the nature of the commercial cooking activities, a single kitchen could have different inspection frequencies. (See Figure 6-16) For example, a restaurant that serves smoked meats must inspect appliances used for the preparation of the meat monthly, whereas other equipment would require inspection every 3 or 6 months, depending on the volume of cooking being performed. [Ref. 609.3.3.1]

You Should Know

Building systems are required to ensure the safe use of a building. The provisions in IFC Chapter 6 not only address their installation but also their operation, maintenance, and testing. Building systems can provide heated or cooled environmental air, a secondary electrical power supply, a means for firefighters to control elevators, a mechanism for abating hazards in a building's electricity distribution system, maintenance of fire-resistive rated construction, and a mechanical ventilation system for the removal of smoke, heat, and byproducts of cooking in commercial kitchens. ●

If the inspection reveals an accumulation of grease, the hood, grease removal devices, exhaust fans and ducts are required to be cleaned. [Ref. 609.3.3.2]

According to the U.S. Fire Administration an estimated 7,100 fires occurred in restaurants in 2002, resulting in an estimated 108 injuries and $116 million in property loss. Although this report contains no estimates of deaths from restaurant fires, the potential for fire fatalities exists in any building or property where people congregate. Based on this loss history, commercial cooking operations present a fire- and life safety threat in many jurisdictions that require close supervision. Accordingly, commercial cooking activities under a Type I hood are required by the IFC to be protected by an automatic fire-extinguishing system. (See Figure 6-17) [Ref. 904.11]

FIGURE 6-16 Cleaning of Type I hoods that serve commercial cooking operations is a required activity to limit the amount of available fuel that could spread to a building

FIGURE 6-17 An automatic fire extinguishing system designed for the protection of commercial kitchen hoods *(Courtesy of Tyco/ Ansul Inc., Marinette, WI)*

Interior Finish and Decorative Materials

The requirements in IFC and IBC Chapter 8 address the selection and installation of materials, fabrics, surface treatments, and furnishings installed or placed inside of buildings. Interior finish are the areas of interior walls, ceilings, and floors and may include fixed or moveable wall partitions and interior wainscoting, paneling, or other finish that is used to decorate, insulate, or to reduce sound levels inside a building. Decorative materials can be natural or man-made materials that are applied over the building's interior finish for decorative or acoustical effects. The interior finish, decorative materials, and furnishings are one source of fuel in a building fire. The requirements in IFC and IBC Chapter 8 were developed after a number of structure fires resulting in hundreds of lives being lost because of exposure to smoke or because the room or area experienced a flashover.

PURPOSE OF THE REQUIREMENTS

The IBC and IFC requirements for interior finishes and furnishing were developed after a number of fires in of assembly occupancies in which highly combustible materials were affixed to the interior ceiling or walls of the building or extremely combustible decorative materials were used that contributed to rapid smoke production and fire growth. A number of major fires in the U.S. and throughout the world have resulted in hundreds of fatalities. Table 7-1 summarizes four of these major incidents. In all of the indicated buildings, the interior finish or decorative materials were a contributing factor—however, other factors including inadequate means of egress width, travel distances, and improper or obstructed exit door openings also contributed to the high number of lives lost.

TABLE 7-1 Summary of large life loss fires where interior finishes or decorative materials were a contributing factor

Location	Date	Number of fatalities	Number of injuries
Rhythm Night Club, Natchez, MS	April 23, 1940	209	Unknown
Cocoanut Grove Night Club, Boston, MA	November 28, 1942	492	Unknown
Beverly Hills Supper Club, Southgate, KY	May 28, 1977	165	> 200
Station Nightclub, Warwick, RI	February 20, 2003	100	> 200

The IFC Chapter 8 requirements limit the likelihood that interior finish or decorative materials contribute to flashover. In Group I and R-2 university dormitories the IFC also prescribes requirements that establish fire safety requirements for furnishings. The requirements limit the amount of heat released when compared to other combustible materials of equal mass and density.

Flashover has a variety of definitions. In the mind of firefighters, a room totally involved in fire or a hot smoke layer that has visible flames are two examples of a flashover. Flashover is an event during a fire's growth where the hot smoke layer inside a room or compartment releases the greatest amount of radiant energy. During the incipient phase of fire growth, carbon monoxide (a flammable gas), and other gases and smoke particulate are produced and begin to accumulate at the top of the room, forming a hot smoke layer.

As the fire continues to progress through the pre-flashover phase, it heats the compartment including furnishings, decorative material, and stored goods, which produces more heated gases and smoke. In the incipient and pre-flashover periods of fire growth, convective heat transfer between the burning materials and the interior surfaces of the room or compartment is the dominating energy. (See Figure 7-1)

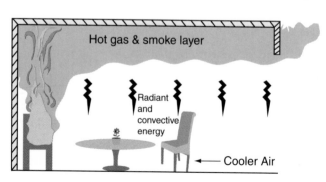

FIGURE 7-1 Pre-flashover conditions in a room or compartment fire

Additional hot gases and smoke are produced as the size of the fire increases. This in turn causes the smoke layer to thicken, to increase in temperature, and to move closer to the combustible fuels. At this point, radiant energy becomes the dominant energy. When the temperature of the smoke layer reaches approximately 600°C (1,100°F), the level of radiant energy absorbed by the combustible materials, causing them to exceed their ignition temperature and flashover occurs. Another criterion of flashover is the radiant energy level at the floor reaches 15–20 kilowatts (kW)/square meter (m^2). These criteria at which flashover is estimated to occur will be due to different mechanisms that result from varying fuel properties, fuel orientation, geometry of the room or enclosure, and conditions in the hot smoke layer.

The likelihood of surviving flashover is almost nil unless the individual is wearing structural firefighting clothing and respiratory protection.

The IFC and IBC provisions limit the likelihood of interior finish and decorative materials contributing to flashover by prescribing the use of flame resistant materials with a minimum density to slow the rate of fire spread and slow the rate of heat release. (See Figure 7-2) The orientation and installation of interior finish materials and decorative materials on room walls and ceilings is also regulated, as these have a profound effect on the time to flashover. Prescribed tests are performed to ensure the material either will not cause a flashover or will have a limited ability to contribute fuel to a flashover, based on the test criterion. Flame resistance also can be provided by a thermal barrier, such as placing gypsum wallboard over expanded foam plastic.

Code Basics

A compartment fire has four phases of growth and decay: incipient, pre-flashover, flashover, and post-flashover. Flashover occurs at approximately 1,100°F and is driven primarily by radiant energy within the fire's hot gas and smoke layer. ●

FIGURE 7-2 Synthetic foam plastic decorations suspended from a building's ceiling. Such a display could accelerate a fire's growth and promote flashover. This material and its installation is a violation of IFC chapter 8

INTERIOR WALL AND CEILING FINISH AND TRIM

The IFC requirements for interior finish materials installed on walls and ceilings are based on its rate of flame spread and smoke production, or its ability to contribute to flashover. Different fire test methods are permitted by the IFC, and each test has its own criteria for assessing the relative fire risks of materials being evaluated. **[Ref. 803.1]**

A fire test that has been used for over 50 years is ASTM E 84, *Standard Test Method for Surface Burning Characteristics of Building Materials.* ASTM E 84 is also known as the Steiner Tunnel Test, named after Edward Steiner, who was an engineer at Underwriters Laboratories who developed the test apparatus. The test involves comparing the performance of the material being evaluated to two materials which have predictable and reproducible test results. The test results are reported as a material's Flame Spread Index (FSI) and Smoke Developed Index (SDI). FSI and SDI are a comparative measure, expressed as a dimensionless number, derived from visual measurements of the spread of flame or smoke obscuration versus time for a material tested in accordance with ASTM E 84 or UL Standard 723. (See Figure 7-3) **[Ref. 802.1]**

Flame spread is assessed visually by the progression of the flame front, while measurements of optical smoke density at the tunnel outlet determine the smoke obscuration. This information is used to plot time-based graphs of flame-spread distance and of optical density. FSI and SDI are then calculated based on the ratio between the duration for the material being tested and those for asbestos cement board (assigned

FIGURE 7-3 ASTM E84 test apparatus for measuring material's flame spread and smoke development indexes *(Courtesy of Underwriter's Laboratories, Northbrook IL)*

FSI and SDI values of 0) and for red oak flooring (assigned FSI and SDI values of 100). Based on the FSI and SDI values, the IFC assigns a Class rating for interior finish materials:

Interior finish material classifications in accordance with ASTM E 84

Material class	Flame spread index	Smoke development index
Class A	0–25	0–450
Class B	26–75	0–450
Class C	76–200	0–450

The class ratings are used to assign minimum interior finish requirements inside buildings. [Ref. 803.1.1]

Although the E84 test provides a reasonable basis for comparing surface flame-spread characteristics of "traditional" building materials, such as wood, the results of this test may not predict actual fire behavior for many contemporary materials, particularly textiles and plastics. Melting and dripping of these materials during the test and the use of a ceiling-mounted sample configuration can skew test results for contemporary materials, discrediting the E84 test results.

Because of these recognized weaknesses with the E84 test, the IFC references a more-current test method, classified as "room-corner tests," which is prescribed by NFPA Standards 286, *Standard Methods of Fire Tests for Evaluating Contribution of Wall and Ceiling Interior Finish to Room Fire Growth*. Tests conducted in accordance with this standard are recognized by the IFC as alternatives to the ASTM E84 approach to flame-spread testing, because room corner tests do a better job of simulating actual fire conditions. In a room corner test, materials are mounted to the walls and/or ceiling of a test room, and the fire exposure is generated by a gas burner that simulates a trash can fire extending to a chair in the corner of the room. (See Figure 7-4) [Ref. 803.1.2]

FIGURE 7-4 A room corner fire test apparatus

FIGURE 7-5 A room corner fire test of a wall and ceiling finish material

The burner flame in the NFPA 286 test contacts the ceiling surface in the latter portion of the test, providing a substantial fire exposure to ceiling-mounted materials. Because the burner flame exposes both the wall and the ceiling in the NFPA 286 test, this test can be used for evaluation of both wall and ceiling finishes. To successfully pass the room-corner test, the material being tested must withstand certain fire exposure and, depending on if the material will be installed on walls, ceilings, or both, the flames must not extend beyond either the walls or boundary of the test compartment. (See Figure 7-5) The acceptance criterion is:

1. During the 40 kilowatt (KW) exposure, flames do not spread to the ceiling.
2. During the 160 KW exposure, the flames do not spread to the outer extremity of the sample on any wall or ceiling and flashover, as defined in NFPA 286, does not occur.
3. The total smoke released throughout the duration of the fire test does not exceed 1,000 square meters. **[Ref. 803.1.2.1]**

Requirements for interior finish materials are set forth in Table 803.3 of the IFC and IBC (see Table 7-2). Table 803.3 establishes these requirements based on the occupancy classification of the building and where the interior finish material will be installed within the means of egress system. Consider a Group A-2 occupancy. If the building is not protected by an automatic sprinkler system, Class A interior finish materials are required in all exit enclosures, exit passageways, and corridors while all rooms or enclosed spaces would require the use of interior finish materials with a Class B rating. If the building is protected throughout by an automatic sprinkler system, the interior finish materials with a higher FSI can be used.

UPHOLSTERED FURNITURE AND MATTRESSES

The IFC has requirements limiting the ignitability and heat release rate of furnishing and bedding in board and care facilities (Group I-1), nursing homes, and hospitals (Group I-2), detention and correctional facilities (Group I-3), and college and university dormitories (Group R-2). The requirements are concerned with furniture and bedding used in areas where individuals sleep or are undergoing patient care and are incapable of self-rescue.

The requirements for determining the ignition resistance of upholstered furniture and mattresses in each occupancy class are based on tests that replicate ignition using a lit cigarette. At the end of the test, the length of the char is measured. If the char length is less than the IFC prescribed limits, which vary for upholstered furniture and occupancies, the furnishing or bedding can be introduced and used in sleeping areas. **[Ref. 805.1.1.1, 805.1.2.1, 805.2.1.1, 805.2.2.1, 805.3.1.1, 805.3.2.1, 805.4.1.1, 805.4.2.1]**

The heat release rate of furnishings in the indicated occupancies is also regulated by the IFC provisions. Furniture inside the indicated

Code Basics

Interior wall and ceiling finish requirements were developed so that selected materials contribute little fuel to a fire. The requirements in the IFC are based on the FSI and SDI of a material. As an alternate to the ASTM E 84 test, the IFC allows the use of a room-corner test, which offers a realistic assessment of a material's fire behavior. •

TABLE 7-2 Interior wall and ceiling finish requirements by occupancy (IFC Table 803.3)

Group	Sprinklered			Nonsprinklered		
	Exit enclosures and exit passageways	Corridors	Rooms and enclosed spaces	Exit enclosures and exit passageways	Corridors	Rooms and enclosed spaces
A-1 & A-2	B	B	C	A	A	B
A-3, A-4, A-5	B	B	C	A	A	C
B, E, M, R-1, R-4	B	C	C	A	B	C
F	C	C	C	B	C	C
H	B	B	C	A	A	B
I-1	B	C	C	A	B	B
I-2	B	B	B	A	A	A
I-3	A	A	C	A	A	B
I-4	B	B	B	A	A	B
R-2	C	C	C	B	B	C
R-3	C	C	C	C	C	C
S	C	C	C	B	B	C
U	No restrictions			No restrictions		

For SI: 1 inch 5 25.4 mm, 1 square foot 5 0.0929m².

a. Class C interior finish materials shall be allowed for wainscoting or paneling of not more than 1,000 square feet of applied surface area in the grade lobby where applied directly to a noncombustible base or over furring strips applied to a noncombustible base and fireblocked as required by Section 803.4 of the International Building Code.

b. In exit enclosures of buildings less than three stories in height of other than Group I-3, Class B interior finish for nonsprinklered buildings and Class C for sprinklered buildings shall be permitted.

c. Requirements for rooms and enclosed spaces shall be based upon spaces enclosed by partitions. Where a fire-resistance rating is required for structural elements, the enclosing partitions shall extend from the floor to the ceiling. Partitions that do not comply with this shall be considered as enclosing spaces and the rooms or spaces on both sides shall be considered as one. In determining the applicable requirements for rooms and enclosed spaces, the specific occupancy thereof shall be the governing factor regardless of the group classification of the building or structure.

d. Lobby areas in Group A-1, A-2, and A-3 occupancies shall not be less than Class B materials.

e. Class C interior finish materials shall be allowed in Group A occupancies with an occupant load of 300 persons or less.

f. In places of religious worship, wood used for ornamental purposes, trusses, paneling, or chancel furnishing shall be allowed.

g. Class B material is required where the building exceeds two stories.

h. Class C interior finish materials shall be allowed in administrative spaces.

i. Class C interior finish materials shall be allowed in rooms with a capacity of four persons or less.

j. Class B materials shall be allowed as wainscoting extending not more than 48 inches above the finished floor in corridors.

k. Finish materials as provided for in other sections of this code.

l. Applies when the vertical exits, exit passageways, corridors, or rooms and spaces are protected by an approved automatic sprinkler system installed in accordance with Section 903.3.1.1 or 903.3.1.2.

occupancies are limited to a maximum peak heat release rate of 80 kilowatts when tested in accordance with ASTM E 1337, *Standard Test Method for Fire Testing of Upholstered Furniture* or using an alternate test method known as California Technical Bulletin 133. Mattresses are limited to a maximum heat release rate of 100 KW. Test samples are also limited to a total energy release of 25 megajoules or less during the first 10 minutes of the test. [**Ref. 805.1.1.2, 805.2.1.2, 805.2.2.2, 805.3.1.2, 805.4.1.2**]

To assist in verifying compliance with the IFC, upholstered furniture and mattresses require a label issued by an approved agency to indicate conformance with the IFC cigarette-ignition resistance and maximum heat release rate requirements. (See Figure 7-6) [**Ref. 805.1.1.3, 805.1.2.3, 805.1.3.3, 805.1.4.3**]

The requirements in Section 805 are waived when the room or area is protected by an automatic sprinkler system designed and installed in accordance with Section 903.3.1.1. In Group I-2 nursing homes and hospitals, the requirements are waived with the patient's room is equipped with a smoke detector or smoke alarm.

FIGURE 7-6 Upholstered furniture and mattresses require a tag or label indicating conformance with the IFC requirements for ignition resistance and maximum heat release rate

PART IV

Fire/Life Safety Systems and Features

Requirements for All Fire Protection Systems

Chapter 9 of IFC sets forth requirements for active fire protection systems. A fire protection system is defined as *approved devices, equipment and systems or combinations of systems used to detect a fire, activate an alarm, extinguish or control a fire, control or manage smoke and products of a fire or any combination thereof.* Given this definition, a fire protection system can include an automatic sprinkler system, an alternative automatic-fire extinguishing system, a fire pump, or smoke alarms in one- and two-family dwellings. A fire protection system prescribed by the IFC must perform one or more functions:

- Detect a fire
- Activate an alarm
- Extinguish or control a fire
- Control or manage smoke or other combustion byproducts of a fire. [**Ref. 902.1**]

Chapter 8 introduces readers to the general provisions in IFC Section 901. These provisions are applicable to all fire protection systems required by the IFC and optional systems that may be specified by the design professional. Installation of fire protection systems that are not required is allowed provided they are installed in accordance with all of the requirements in the IFC and IBC. [Ref. 901.4.2]

WHEN ARE FIRE PROTECTION SYSTEMS REQUIRED?

The provisions requiring a fire protection system vary in the IFC and IBC. Outside of buildings, the IFC requires the installation of private fire protection water mains and fire hydrants to ensure that the required fire flow is available in the event of a structure fire. IBC and IFC fire protection system requirements inside of buildings are based on:

1. The occupancy of the building or its fire area.
2. The occupant load of the building or its occupancy.
3. The height or area of a building.
4. The quantity or hazards of materials stored or used inside a building.
5. The fire loss history of a given hazard. [Ref. 901.4.1]

Certain occupancy classifications require the installation of one or more fire protection systems based on these five characteristics. In the case of a Residential occupancy (Group R), an automatic sprinkler system is required throughout the building when a Group R fire area is created. For example, if part of a small one-story Business (Group B) were converted to a motel, this change of occupancy requires the installation of an automatic sprinkler system throughout the entire building. Conversely, the construction of a moderate hazard factory occupancy (Group F-1) would not require automatic sprinkler protection unless its fire area exceeds 12,000 square feet. Many of the IFC fire protection system requirements are based on the building occupancy, occupant load, or its fire area. [Ref. 903.2.8, 903.2.4]

The occupant load of a building will dictate the installation of certain fire protection systems. In an Assembly occupancy, such as a dance hall where food or beverages are not served (Group A-3), an automatic sprinkler system is required if the occupant load exceeds 300 persons. Because of the large occupant load, the IFC prescribes the installation of occupant notification devices as part of a manual fire alarm and detection system that would inform occupants in the event that the automatic sprinkler system operates. (See Figure 8-1) [Ref. 903.2.1.3, 907.2.1]

A building's height or area can dictate the installation of a fire protection system. The IBC limits building height and area based on its occupancy classification and construction type. (See Figure 8-2) IBC construction types are fire-resistive, noncombustible, combustible, or a mixture of combustible and noncombustible materials. In many occupancies, the IBC permits the building's height and area to be increased

FIGURE 8-1 A Group A-3 occupancy such as a live music venue in an airport would require automatic sprinkler protection and a fire alarm system with occupant notification throughout the fire area

FIGURE 8-2 The IBC occupancy and building's construction type classification establish the allowable height and area. Allowable height and area can be increased by certain designs of automatic sprinkler protection

when the building is protected by an automatic sprinkler system. The number and type of increases granted depends on if the automatic sprinkler system is designed for property and life safety protection or if it is only designed for life safety applications. [Ref. 903.1.1]

A fire protection system may be required inside a building when the quantity of materials stored or their hazards exceed certain limits. IFC Chapter 23 regulates High-Piled Combustible Storage, which are very common methods of storing large volumes of materials in Group S occupancies. (See Figure 8-3) Automatic sprinkler protection in these buildings is required based on the fire hazard of the stored goods, the storage height and area, and the building's occupancy classification. Certain classes of hazardous materials are so easily ignited or can cause such a large amount of damage if ignited that IFC Chapter 27 requires the installation of an automatic sprinkler system throughout buildings when these classes of hazardous materials are stored indoors. [Ref. 903.2.7.1, Table 2703.1.1(1) footnote g]

Certain hazards inside of buildings represent a relatively high threat to the occupants, the building, and its content if an unwanted fire occurs. (See Figure 8-4) To control the fire risks, the IFC requires the installation of fire protection systems for specific hazards or uses. One example is commercial cooking systems that can generate smoke and grease-laden vapors. Because the smoke and vapors act as fuel inside of the commercial cooking exhaust system, the IFC requires the installation of a wet-chemical or dry-chemical alternative fire-extinguishing system. The system is designed in accordance with the applicable National Fire Protection Association (NFPA) standards and IFC installation requirements. Any modification to an existing commercial cooking system requires the fire protection system be upgraded to comply with the IFC. [Ref. 904.11 and 904.11.6.1]

FIGURE 8-3 The quantity of stored materials or their hazards can mandate the installation of a fire protection system such as High-Piled Combustible Storage, as shown in this photograph

FIGURE 8-4 The IFC requires certain fire hazards that represent a relatively high threat to building occupants, contents, and the structure itself be protected by a fire protection system *(Courtesy of Ansul/Tyco Inc., Marinette WI)*

A jurisdiction may be asked to approve the construction of a building, structure, or process where the hazards are special or challenging to emergency responders. In some cases, the size or arrangement of the hazard may impair or limit the ability of fire apparatus to approach to location. In such instances, the fire code official can require additional fire protection safeguards. These can be in the form of any fire protection system as defined in Section 902. Consistency in code enforcement is important. Once the fire code official decides additional protection is required, it must be also required for any similar installations in the future—if not, the jurisdiction could be liable for the added expenses imposed on one business while waiving the requirement for others.

Additional fire protection systems or safeguards may not be so obvious or rely on fire suppression and detection systems. (See Figure 8-5) Realize that certain process safety controls specified by the design professional, such as installing certain flow control valves, automatic shut-down components, or other less-known but equally-reliable systems, may effectively reduce and manage a hazard while providing greater reliability than a fire suppression or detection system. In instances when the fire code official is considering additional fire protection systems, it's reasonable and prudent to request that the design be reviewed by a fire protection engineer or other competent design professional who has the experience and understanding of the hazards of the building, occupancy, the stored materials or the process. (See Chapter 1) **[Ref. 901.4.3]**

> **Code Basics**
>
> The IFC requires fire protection systems inside and outside of buildings. When a fire protection system is required indoors, it can be based on the building's occupancy classification, occupant load, its height or area, or the hazards of the stored goods and materials. •

CONSTRUCTION DOCUMENTS AND ACCEPTANCE TESTING

IFC Section 105.7 requires a construction permit for the installation or a modification of a fire protection system. Fire protection systems design includes compliance with the adopted NFPA standards for the specific

FIGURE 8-5 This additional fire protection system can be required by the fire code official. The manual master stream nozzles are used for fire exposure protection of large aboveground storage tanks storing aviation jet fuel

systems. Table 8-1 lists the fire protection systems and standards adopted by the IFC. **[Ref. 901.6.1]**

The IFC commonly requires more than one fire protection system in certain buildings. A building with an occupied floor more than 75 feet above the lowest level of fire department access is defined as a High-Rise building by the IBC and requires not only an automatic sprinkler system but also a

TABLE 8-1 Fire protection systems and standards

Fire protection system or component	NFPA standard
Low, Medium, and High Expansion Foam	11
Carbon Dioxide Extinguishing System	12
Halon 1301 Fire Extinguishing System	12A
Installation of Sprinkler Systems	13
Installation of Sprinkler Systems in One- and Two-Family Dwellings and Manufactured Homes	13D
Installation of Sprinkler Systems in Residential Occupancies up to and Including Four Stories in Height	13R
Installation of Foam-Water Sprinkler and Foam-Water Spray Systems	16
Dry Chemical Extinguishing Systems	17
Wet Chemical Extinguishing Systems	17A
Installation of Stationary Pumps for Fire Protection	20
Water Tanks for Private Fire Protection	22
Installation of Private Fire Service Mans and Their Appurtenances	24
National Electrical Code©	70
National Fire Alarm Code	72
Water Mist Fire Protection Systems	750
Water Supply for Suburban and Rural FireFighting	1142
Clean Agent Fire Extinguishing System	2001

standpipe system, a fire alarm and detection system equipped with an emergency voice/alarm communication system, a generator to provide standby and emergency power, and possibly a fire pump. All of these systems must comply with the IFC and NFPA requirements. [**Ref. 901.4.1**]

The IFC requires the permit applicant to submit construction documents to the fire code official for review and approval of the fire protection system to verify the installation or modification is properly designed for the hazards it is protecting. (See Figure 8-6) The submittal scope and detail is commonly specified in the NFPA standard applicable to the system design or modification. In many cases the submittal will include manufacturer's equipment data or cut sheets and various formats of engineering calculations used as the basis for the design. [**Ref. 901.2 and 907.1.1**]

Prior to requesting a final approval of a fire protection system installation or modification, the fire code official can require the installing contractor to submit a Statement of Compliance. The statement of compliance documents that the fire protection system was installed in accordance with the applicable NFPA installation standard, the construction specification prepared by the design professional, and the manufacturer's instructions for particular components. Any deviations from the installing standard are also attached to the Statement of Compliance. This statement may also be used to document that the fire protection system has been tested by the installing contractor—this is an important consideration, especially on larger or complicated systems. If the fire protection system is not tested and exercised prior to a final inspection, it generally fails to pass the inspection. [**Ref. 901.2.1**]

The NFPA standards that govern the design of automatic sprinkler systems, wet-chemical alternative fire-extinguishing systems, private underground fire protection water mains, and fire alarm and detection systems

FIGURE 8-6 Fire protection system construction documents must be submitted to the fire code official to verify that the design will protect the identified hazards and will meet the requirements of the applicable technical standards

each contain an example Statement of Compliance. (See Figure 8-7) In instances where an example statement of compliance is not available, the jurisdiction can develop its own, require the installing contractor to provide such a statement, waive the requirement when allowed by the fire code official, or use example forms available in the *Fire Plan Review and Inspection Guidelines* published by ICC. In states that license contractors, a Statement of Compliance may be required to be prepared by the installing contractor.

INSPECTION, TESTING, AND MAINTENANCE

Fire protection systems require mechanical and electrical components to properly operate and to perform their intended functions. Water-based fire protection systems must be connected to a reliable source of water and

Stationary Fire Pump Statement of Compliance

Permit #:_____ Date: _____

	Property Protected	Installing Contractor	General Contractor
Business Name:			
Address:			
Representative:			
Telephone:			

Location of Plans: _____

Location of Owner's Manual: _____

1. Certification of System Installation: This system installation was inspected and found to comply with the installation requirements of:

_____ NFPA 20 and the National Electric Code©
_____ Manufacturer's Instructions
_____ Other (specify; FM, UL, etc.) _____

Print Name: _____

Signed: _____ Date: _____

Organization: _____

2. Certification of System Operation: All operational features and functions of this system were tested and found to be operating properly in accordance with the requirements of:

_____ NFPA 20 and National Electric Code©
_____ Design Specifications and Submitted Ship Drawings
_____ Manufacturer's Instructions
_____ Other (specify) _____

Print Name: _____

Signed: _____ Date: _____

Organization: _____

FIGURE 8-7 Stationary fire pump of statement of compliance

alternative automatic-fire extinguishing systems require a sufficient volume of wet, dry, or gaseous agent to extinguish a fire. Fire protection systems that discharge water or another alternative agent are constructed with listed nozzles that must not be obstructed, blocked, painted, or improperly oriented. Fire protection systems must be maintained so they are available when ignition occurs. Defective components that have failed or that are inoperable must be repaired or replaced. Non-required fire protection systems also must be inspected, tested, and maintained. [Ref. 901.6]

TABLE 8-2 Fire protection system inspection, testing, and maintenance

Inspection	A periodic visual check to assure the system or equipment is in place, is not impaired, and appears ready for operation.
Testing	A functional or performance test performed by a qualified person to verify that the system or equipment will operate as intended.
Maintenance	Periodic service performed by a qualified person in accordance with the system or equipment manufacturer's recommendations to extend operational life and to enhance reliability.

Required and non-required fire protection systems must be inspected, tested, and maintained in accordance with the applicable fire protection system standards (Table 8-2). For water-based fire protection systems including private water storage tanks, fire hydrants supplied from private fire protection water mains, automatic sprinkler systems, standpipes, and fire pumps, the IFC adopts NFPA 25, *Standard for the Inspection, Testing, and Maintenance of Water-Based Fire Protection Systems*. Most water-based fire protection systems require an annual inspection and one or more tests. The IFC prescribes increased inspections intervals for alternative automatic-fire extinguishing systems. NFPA standards for inspection, testing, and maintenance are set forth in Table 901.6.1 (see Table 8-3). [Ref. 901.6.1]

Inspection, test, and maintenance records must be maintained on the premises for at least three years. (See Figure 8-8) Upon request, these records must be copied to the fire code official. In addition, the initial

TABLE 8-3 Fire protection system maintenance standards (IFC Table 901.6.1)

System	Standard
Portable fire extinguishers	NFPA 10
Carbon dioxide fire-extinguishing systems	NFPA 12
Halon 1301 fire-extinguishing systems	NFPA 12A
Dry-chemical extinguishing systems	NFPA 17
Wet-chemical extinguishing systems	NFPA 17A
Water-based fire protection systems	NFPA 25
Fire alarm systems	NFPA 72
Water-mist systems	NFPA 750
Clean-agent extinguishing systems	NFPA 2001

FIGURE 8-8 The IFC requires fire protection systems be inspected, tested, and maintained

FIGURE 8-9 A wet pipe automatic sprinkler system with a closed water supply valve. Impairments can be identified by the fire protection water supply, which must indicate the valve's state (open or closed) and pressure gages. If a pressure gauge reads zero, the system is most likely unavailable

installation records for the fire protection system must be maintained at the site. The records must include the name of the installing contractor, equipment cut and data sheets, and the manufacturer's installation and instruction manuals. **[Ref. 901.6.2]**

FIRE PROTECTION SYSTEM IMPAIRMENT

When a fire protection system is impaired, the fire department and fire code official must be notified. When a system is out of service, the fire code official can either require the evacuation of the building or permit the continued use of the building under the supervision of a fire watch. After the fire protection system is returned to service, the fire watch is discontinued. **[Ref. 901.7]**

Impairments of fire protection systems can be categorized into two broad categories: unscheduled or scheduled. An unscheduled impairment includes the loss of utility electricity to an electric fire pump or the failure of a water main supplying an automatic sprinkler system. An unscheduled impairment could be an individual turning off a water supply valve to a sprinkler system before committing criminal arson. A scheduled impairment includes testing, maintenance, or modifications to a fire protection system. (See Figure 8-9) For example, if a tenant improvement is constructed in a covered mall building where several lease spaces are consolidated into a single retail space, automatic sprinkler system modification is an anticipated and foreseeable event. When a fire protection system is impaired, either the building owner or a designated employee assumes the role of impairment coordinator. An impairment coordinator is also responsible for the maintenance of the fire protection system. **[Ref. 901.7.1]**

The impairment coordinator must perform several actions before authorization can be granted to remove the system from service. These actions include:

1. Identify the extent and duration of the impairment.
2. Inspect the areas of the building that will be affected by the impairment and identify any processes or hazards that need to be discontinued. In some cases, this may require prescribing the use of less hazardous operations such as using hand tools rather than machinery.
3. Notify the fire department, the approved fire alarm system monitoring station, as well as the required notifications within the organization's hierarchy, such as the loss control department, the safety coordinator, and the insurance underwriter.
4. Assemble the necessary tools and materials to perform the modification or impairment.
5. Implement the impairment tag program. [Ref. 901.7.4]

After the impairment is concluded, the impairment coordinator must notify the organizations and individuals specified in Section 901.7.4 and inspect and test the fire protection system to verify it is operational and available for service. [Ref. 901.7.6]

When impairment occurs, a tag is required at specified locations to indicate the fire protection system is out of service. (See Figures 8-10A and 8-10B) These locations include the fire department connection, system control valve(s), fire alarm control unit, fire alarm annunciator, and, if one exists, the fire command center. Once the system has been returned to service, the impairment tag is removed. [Ref. 901.7.2, 901.7.3]

FIGURE 8-10A AND 8-10B An impairment tag is required to be posted at specific locations

No.: 11500

Fire Protection System or Component
Impairment
Removal of this tag shall be authorized by the Impairment Coordinator

Address: _____

Building No.: _____   Floor Level: _____

Fire Protection System Impaired

☐ Automatic Wet Pipe Sprinkler System - System Number _____

☐ Automatic Dry Pipe Sprinkler System - System Number _____

☐ Automatic Pre-Action Sprinkler System - System Number _____

☐ Alternative Automatic Fire Extinguishing System - System Number _____

☐ Automatic Fire Alarm System - System Number _____

☐ Kitchen Hood Fire Suppression System - System Number _____

☐ 1,500 GPM Electric Fire Pump   ☐   Fire Door - Number _____

Impairment Information

Description of Problem: _____

Reported By: _____ DATE: ☐☐☐☐☐☐

Restoration to Service

Summary of Repairs: _____

Repaired By: _____ DATE: ☐☐☐☐☐☐

☐ 2-inch Main Drain Flow Test Performed Static P _____ Residual P _____

☐ Alarm Initiating Device Test and Alarm Signal Verification

Impairment Coordinator Verification

Observed By: _____ DATE: ☐☐☐☐☐☐

☐ Restoration Approved   ☐ Restoration Not Approved

FIGURE 8-10A AND 8-10B (Continued) An impairment tag is required to be posted at specific locations

FIRE PROTECTION SYSTEM MONITORING

In most occupancies, the IFC requires monitoring of automatic sprinkler systems and fire alarm and detection systems. Monitoring includes detecting and reporting alarm conditions and potential impairments, such as closed water supply valves or electrical faults, and is required to ensure a timely response by the fire department. Monitoring is not required for automatic sprinkler systems protecting one- and two-family dwellings and townhouses required by the IRC provisions. [Ref. 903.4]

Automatic sprinkler system monitoring is electric supervision of valves that control the water supply to the system and water flow alarms or pressure switches that operate when a sprinkler is activated. (See Figure 8-11) Upon activation, these devices transmit a signal to a fire alarm control unit. The fire alarm control unit in turn transmits a signal by way of telephony, the Internet, or through a wireless signal to a central or proprietary supervising station. A central supervising station is a third-party service that receives signals from fire and security systems, processes the signals, and notifies the fire department and building owner of an alarm activation. A proprietary supervisory station is one owned by the property being protected, such as a campus police department. [Ref. 903.4.1]

With the exception of smoke alarms in one- and two-family dwellings and correctional and detention facilities (Group I-3 occupancies), all required fire alarm and detection systems required by the IFC must be electrically supervised. [Ref. 907.7.5]

You Should Know

The intent of the IFC fire protection system requirements:

- It is designed and constructed in accordance with the applicable NFPA standards.
- It is designed to protect the hazards inside the building.
- Signals from most required automatic sprinkler and fire alarm and detection systems are transmitted to a central monitoring station, which in turn notifies the fire department.
- The system is inspected, tested, and maintained in accordance with the IFC and the applicable NFPA standards. •

FIGURE 8-11 Electric supervision of an indicating floor control valve and water flow switch inside a high rise building

Automatic Sprinkler Systems

Automatic sprinkler systems are the most reliable fire protection system. They are designed to detect, report, and control a fire until the fire department arrives to suppress it. An automatic sprinkler system provides a means of fire detection, because sprinklers are constructed with a heat sensitive element that operates within a specific temperature range during the early stage of fire growth. When heated above a preset temperature, the sprinkler fusible link or frangible bulb operates releasing a controlled volume of water directly onto the fire. The discharged water absorbs the heat produced by a fire and cools the air in the room or around involved the fire reducing the amount of heat released and slowing its rate of fire spread. Activation of the sprinkler system will transmit an electric signal to a supervising station which in turn notifies the fire department.

The IBC and IFC offer many credits for buildings protected by an automatic sprinkler system. Under the IBC height and area provisions for certain occupancies, a building's height can be increased by one story and the allowable area increased up to 300% when it is protected by an automatic sprinkler system. The IFC allows a site's fire flow to be reduced by 75% when buildings are sprinklered. These are three of the many fire protection trade-offs allowed by the I-codes. Because many of these trade-offs can influence the ability of the fire department to control a fire if the automatic sprinkler system does not perform its function, it is important that the design and installation of automatic sprinkler system comply with the applicable IFC and NFPA requirements.

LEVEL OF EXIT DISCHARGE AND FIRE AREA

When applying the IFC and IBC automatic sprinkler systems requirements, two prerequisite terms must be understood: Level of Exit Discharge and Fire Area. Proper application of these two terms, along with the building occupancy, is essential in determining when an automatic sprinkler system is required.

The Level of Exit Discharge is defined as *the story at the point at which an exit terminates and an exit discharge begins.* (See Figure 9-1) The level of exit discharge is the last portion of the egress system and is the portion where an occupant leaves as exit and continues until a public way is reached. All travel that is outside the building is considered as part of the exit discharge until the person reaches the public way. At this point, exiting is complete and the person is deemed safe under the means of egress provision in the IFC. A building can have more than one level of exit discharge. **[Ref. 1002.1]**

A building's fire area is *the aggregate floor area enclosed and bounded by fire walls, fire barriers, exterior walls, or horizontal assemblies of a building. Areas of the building not provided with surrounding walls shall be included in the fire area if such areas are included within the horizontal projection of the*

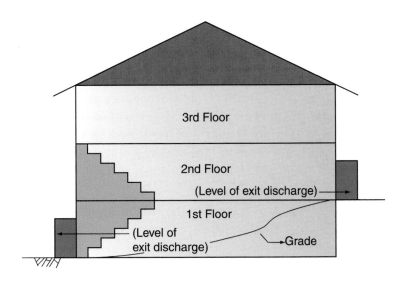

FIGURE 9-1 Level of exit discharge

Given: A 18,000 square foot building with two tenants: A 11,000 square foot group M retail store and a 7,000 square foot group S-1 storage occupancy without high-piled combustible storage.

Determine: The requirements for using fire areas rather than an automatic sprinker system.

Solution: Group M fire area ≤ 12,000 Ft.^{2}
Group F-1 fire area ≤ 12,000 Ft.^{2}
Total of group M and F-1 fire areas ≤ 24,000 Ft.^{2}

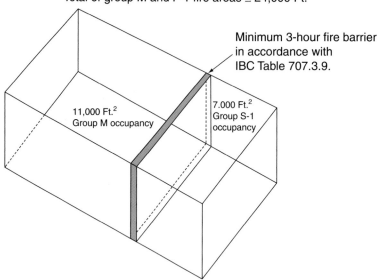

FIGURE 9-2 Example application of fire areas

roof or floor next above. (See Figure 9-2) A fire area is means of creating a compartment of a given area to limit the spread of fire in a building. By using fire-resistive construction, compartments are created to contain the spread of fire for period of time. By definition, a fire area is the aggregate floor area enclosed and bound by fire walls, fire walls, exterior walls, or fire-resistive horizontal assemblies of a building. By separating a fire area using fire-resistive construction, only a portion of a building is considered at risk during a fire incident. A building of combustible or noncombustible construction can be a fire area. **[Ref. 902.1]**

The fire area methodology set forth in the IBC, applicable only in limited occupancy groups under limited conditions, allows for the omission of automatic sprinkler protection.[1] When a fire wall or fire barrier is used to compartmentalize or separate a structure, only the fire area that exceeds the IFC area limits for the occupancy requires automatic sprinkler protection. Using fire resistive construction to create multiple fire areas in a single building is a design alternative to the requirements for an automatic sprinkler system. Other fire protection system requirements, such as fire alarm and detection systems, are not prescribed using the fire area method. When a fire wall or fire barrier is used to compartmentalize or separate a structure, only the fire area that exceeds the IFC area limits for the occupancy requires automatic sprinkler protection.

Code Basics

The IFC and IBC requirements for an automatic sprinkler system are based on the occupancy and its:
- Fire area or building area
- Location in relation to level of exit discharge
- The occupant load of the fire area
- The amount of combustible or hazardous materials ●

---

[1]Thornburg, Douglas, *2006 IBC Handbook – Fire and Life Safety Provisions*, International Code Council, Washington D.C., pg. 250.

DESIGN AND INSTALLATION STANDARDS

The IFC adopts three NFPA standards that address the design and installation of automatic sprinkler systems. The standards governing the design of the automatic system required by the IFC for the protection of occupancies or buildings will influence the design of the building. The adopted NFPA standards IFC are:

- NFPA 13, *Installation of Sprinkler Systems*
- NFPA 13R, *Sprinkler Systems for Residential Occupancies Up to and Including Four Stories in Height*
- NFPA 13D, *Sprinkler Systems for One- and Two-Family Dwellings and Manufactured Homes*

Automatic sprinkler systems for life safety and property protection are designed in accordance with the requirements of NFPA 13. Depending on the occupancy, automatic sprinkler systems used for life safety are designed to either NFPA 13R or NFPA 13D. NFPA 13R systems are designed to protect Group R-1 and R-2 occupancies up to four stories in height while NFPA 13D is used to design these systems in Group R-3 and R-4 occupancies. In one- and two-family dwellings and townhomes, NFPA 13D can be used as an alternate to the requirements in Section P2904 of the *International Residential Code for One- and Two-Family Dwellings and Townhomes* (IRC) for dwelling fire sprinkler systems. **[Ref. 903.3.1.1, 903.3.1.2, 903.3.1.3]**

The three NFPA standards have their own specific design requirements for the type of sprinklers; water supplies and extent of protection will vary. Table 9-1 summarizes these major considerations for each of the three standards and IRC Section P2904.

TABLE 9-1 Automatic Sprinkler System Design Considerations

Design Consideration	Sprinkler Standard		
	NFPA 13	NFPA 13R	IRC Section P2904 or NFPA 13D
Extent of Protection	Throughout the entire building (Section 903.3.1.1)	Occupied spaces (Section 903.3.1.2)	Occupied spaces (Section 903.3.1.3)
Design Intent	Life Safety and Property Protection	Life Safety	Life Safety
Applicability	All occupancies	Group R occupancies up to 4 stories	One- and two-family dwellings and townhomes up to 3 stories
Design Methods	Pipe schedule Control mode – density/design area Control mode – specific application Suppression modes	4 sprinklers in the hydraulic remote compartment	Prescriptive design using IRC Section P2904; 2 sprinklers in the hydraulic remote compartment using NFPA 13D
Sprinklers	All listed and approved sprinklers	Listed residential sprinklers	Listed residential sprinklers
Minimum Water Supply Duration	30 to 120 minutes depending on the hazard and design	30 minutes	IRC Section P2904: Between 7 and 10 minutes depending on dwelling area and the number of stories; NFPA 13D: 10 Minutes

With the exception of the IRC Section P2904 prescriptive design and the pipe schedule design methods, all other design methods are prepared using hydraulic calculations. Pipe schedule systems, which do not require hydraulic calculation, are only allowed for light- and ordinary-hazard sprinkler designs. Extra hazard sprinkler designs cannot be prepared using the pipe schedule method. Hydraulic calculations that prove the water supply is capable of providing the required pressure and volume of water to the sprinkler system must be a part of the plan review package submitted to the fire code official. Requirements for the preparation of these calculations are established in NFPA 13. Calculations are based on the available pressure and flow rate of the water supply, changes in elevation between the water supply and the sprinklers, the selected sprinklers, the NFPA 13 hazard classification for the building, the loss of pressure that results from the friction of water moving inside the pipe. Friction loss is influenced by the flow rate, the material of construction and the pipe diameter, length and the number and type of connected fittings.

Automatic sprinkler systems are designed to either control or suppress a fire. Many sprinkler systems are designed using the control mode–density/ design area method, which is a design intended to control the fire. Fire control is accomplished when a sufficient volume of water is discharged to control the fire and begin prewetting of exposure commodities. Prewetting slows the spread of fire. The control-mode method relies on the application of a sufficient amount of water in the right location to lower fire gas temperatures so the rate of fire growth is slowed. (See Figure 9-3)

Control mode–specific application automatic sprinklers have a larger discharge orifice when compared to sprinklers designed for density/ design area applications. These sprinklers are designed to:

1. prewet combustibles surrounding the actual fire area to limit fire size,
2. cool the roof area directly over the fire to prevent structural failure, and
3. cool the roof area remote from the actual fire area so that too many sprinklers do not open and overtax the water supply.

FIGURE 9-3 Control mode sprinkler *(Courtesy of TYCO Fire Suppression and Building Products, Lansdale, PA)*

These sprinklers differ from control mode–density/area sprinklers in that a fire protection system utilizing control mode–specific application sprinklers is designed based on a given number of sprinklers operating at a specified minimum water supply pressure. Control mode-specific application automatic sprinklers offer the advantage of higher discharge densities at lower water supply pressures, which can achieve adequate sprinkler protection without requiring a fire pump.

Suppression designs commonly involve the use of Early Suppression Fast Response (ESFR) sprinklers. The design of these sprinklers requires that they suppress (extinguish) the fire. ESFR sprinklers bring a higher level of protection to buildings, however they are intended for very specific uses and have very restrictive rules for installation. These installation rules must be respected over the life of the building. The performance of ESFR sprinklers can be seriously impacted by obstructions such as light fixtures or banners hanging from the structure. These sprinklers require

Code Basics

Residential sprinkler systems are designed to control a fire and prevent flashover. Sprinkler systems in commercial buildings are designed to either control or suppress a fire, depending on their design, and will also prevent flashover. •

constant evaluation, especially in speculation warehouses. Fire code officials should be cautious of building owners or design professionals who believe that ESFR sprinklers will protect all buildings from all hazards.

The expression "suppression" relates to sprinkler system performance whereby the first few sprinklers to operate provide sufficient water to the fire to reduce it promptly to an acceptable level, if not extinguish it. The effectiveness of suppression mode sprinklers depends on the combination of fast response and the quality and efficiency of the sprinkler discharge. (See Figure 9-4)

FIGURE 9-4 Suppression mode sprinkler *(Courtesy of TYCO Fire Suppression and Building Products, Lansdale, PA)*

TYPES OF AUTOMATIC SPRINKLER SYSTEMS

Automatic sprinkler systems are designed to discharge a given volume of water over the area being protected. Their features can vary because of the hazards of materials being stored, the lack of heat to maintain warm air inside a building so the pipe does not freeze, or the potential risk of property damage resulting from accidental operation of the sprinklers.

NFPA 13 establishes design and installation requirements for four different types of automatic sprinkler systems:

- Wet-pipe automatic sprinkler system
- Dry-pipe automatic sprinkler system
- Pre-action automatic sprinkler system
- Deluge automatic sprinkler system

Wet pipe automatic sprinkler systems have the most reliable design because they require the least number of components. (See Figure 9-5 and 9-5A) The piping supplying the sprinklers is charged with water—upon sprinkler activation, water is immediately discharged onto the fire. These systems are permitted in any building where the temperature is maintained at 40°F or more.

Dry pipe automatic sprinkler systems are used in any environment where the temperature is less than 40°F, including loading docks, attics, and refrigerated storage warehouses. (See Figure 9-6) The piping above the dry pipe alarm valve is charged with compressed air or nitrogen. The air or nitrogen forces the alarm valve closed so water cannot enter and freeze inside the pipe. When the sprinkler activates, it exhausts air through its discharge orifice, which lowers the pressure inside the pipe, allowing the alarm valve to open and begin flowing water toward the open sprinkler. Depending on the occupancy being protected, NFPA 13 requires the discharge of water from the open sprinkler within 15 to 60 seconds of operation. On many dry pipe systems, a device called an accelerator, exhauster, or quick-open device is installed to help accelerate the exhausting of the air or nitrogen.

A pre-action automatic sprinkler system utilizes a supplemental fire detection system located in the same area as the sprinklers. (See Figure 9-7) Depending on the design, the piping may be charged with compressed air. The fire detection system activates the alarm

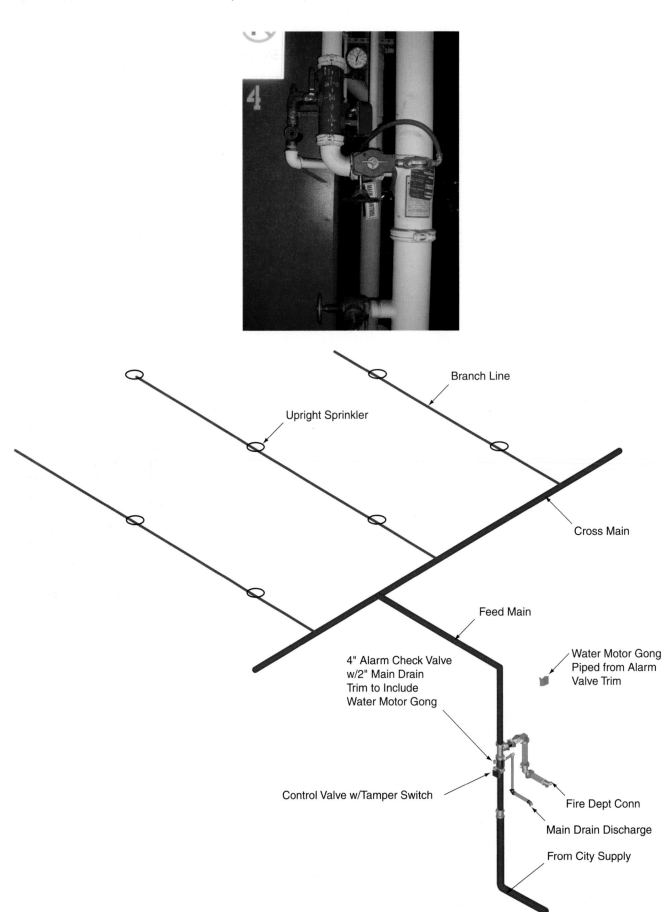

Branch Line

Upright Sprinkler

Cross Main

Feed Main

Water Motor Gong
Piped from Alarm
Valve Trim

4" Alarm Check Valve
w/2" Main Drain
Trim to Include
Water Motor Gong

Control Valve w/Tamper Switch

Fire Dept Conn

Main Drain Discharge

From City Supply

FIGURE 9-5 and 9-5A Wet-pipe automatic sprinkler system *(Courtesy of MFP Fire Protection Design, Gilbert, AZ)*

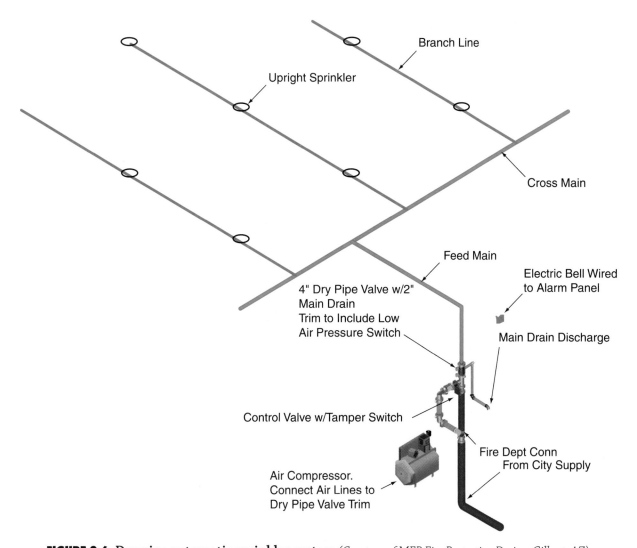

FIGURE 9-6 Dry pipe automatic sprinkler system *(Courtesy of MFP Fire Protection Design, Gilbert, AZ)*

valve, allowing the pipe to be filled with water. Such a design is known as a single-interlock pre-action system. A pre-action sprinkler that only admits water into the pipe upon operation of the sprinkler and the fire detection system is termed a double-interlock system. NFPA 13 does not allow more than 1,000 sprinklers to be controlled by a one single-interlock alarm valve. Single- and double-interlock pre-action sprinklers must discharge water within 60 seconds of the sprinkler's operation.

Deluge automatic sprinkler systems are similar to pre-action sprinklers because they are also activated by a fire detection system located in the same area as the sprinklers. (See Figure 9-8) The sprinklers in this fire protection system are open—the heat detection element is removed. When a deluge sprinkler system operates, water flows from all of the sprinklers. Because all of the sprinklers flow water upon activation, the design of these systems is limited to the available water supply. Deluge automatic sprinkler systems are generally limited to the protection of very challenging goods or materials such as acetylene gas cylinder transfilling plants or storage of certain flammable liquids in plastic packaging.

Code Basics

NFPA 13 sets forth design requirements for four different types of automatic sprinkler systems:
- Wet-pipe
- Dry-pipe
- Pre-action
- Deluge ●

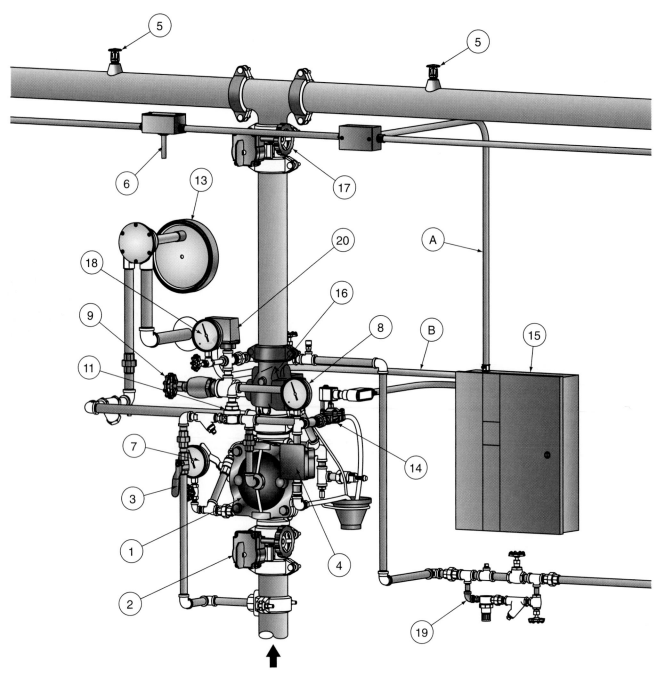

1 - Model DV-5 Deluge Valve	9 - System Drain Valve (N.C.)	16 - Riser Check Valve
2 - Main Control Valve (N.O.)	10 - Main Drain Valve (N.C.)	17 - System Shut-Off Valve (N.O.)
3 - Diaphragm Chamber Supply	(Shown at Rear of Valve)	18 - Air Pressure Gauge
Control Valve (N.O.)	11 - Diaphragm Chamber Automatic	19 - Automatic Air/Nitrogen Supply
4 - Local Manual Control Station	Shut-Off Valve	20 - Low Pressure Alarm Switch
5 - Automatic Sprinklers	12 - Waterflow Pressure Alarm Switch	A - Fire Detection Initiating Circuit
6 - Heat Detectors, Smoke Detectors,	(Shown at Rear of Valve)	(Zone 1)
etc. (Fire Detection)	13 - Water Motor Alarm (Optional)	B - Low Pressure Alarm Initiating Circuit
7 - Water Supply Pressure Gauge	14 - Solenoid Valve	(Zone 2)
8 - Diaphragm Chamber Pressure	15 - Cross-Zone Deluge Valve	
Gauge	Releasing Panel	

FIGURE 9-7 **Pre-action automatic sprinkler system** *(Courtesy of TYCO Fire Suppression and Building Products, Lansdale, PA)*

(Note: The term "N.C." means "normally closed.")

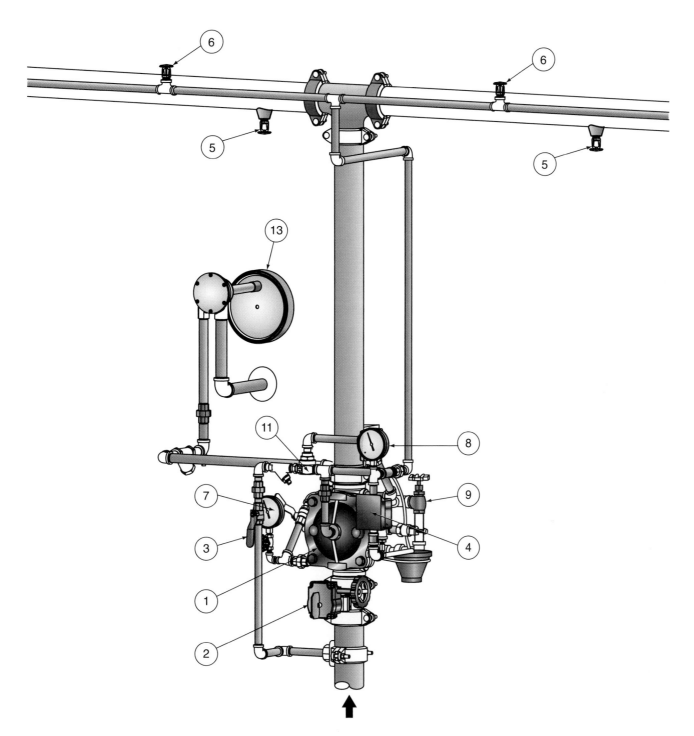

1 - Model DV-5 Deluge Valve
2 - Main Control Valve (N.O.)
3 - Diaphragm Chamber Supply
    Control Valve (N.O.)
4 - Local Manual Control Station
5 - Open Nozzles or Sprinkles
6 - Wet Pilot Line Sprinklers
    (Fire Detection)

7 - Water Supply Pressure Gauge
8 - Diaphragm Chamber Pressure
    Gauge
9 - System Drain Valve (N.C.)
10 - Main Drain Valve (N.C.)
    (Shown at Rear of Valve)
11 - Diaphragm Chamber Automatic
    Shut-Off Valve

12 - Waterflow Pressure Alarm Switch
    (Shown at Rear of Valve)

13 - Water Motor Alarm (Optional)

FIGURE 9-8 Deluge automatic sprinkler system *(Courtesy of Tyco Fire Suppression and Building Products, Lansdale PA)*

(Note: The term "N.C." means "normally closed.")

OCCUPANCIES REQUIRING AUTOMATIC SPRINKLER PROTECTION

Code requirements for automatic sprinklers can be found among the variety of IBC occupancy classes. Buildings housing a Group H-5, I or R fire area require the installation of automatic sprinkler protection throughout the building due to the life safety and fire protection risks. In the case of Group I occupancies (hospitals, nursing homes, and penitentiaries), the occupants are not capable of self-rescue. For Group R occupancies the concern is the ability of the person to be awakened and perform self-rescue of themselves and any other individuals in the building. Buildings housing Group H-5 occupancies (semiconductor fabrication facilities) require automatic sprinkler protection throughout because of the amount and variety of hazardous materials. [Ref. 903.2.5.2, 903.2.6, and 903.2.8]

In assembly occupancies, the requirements for when automatic sprinkler protection are based on the occupancy's fire area or occupant load. In Group A-1, A-3, and A-4 occupancies, automatic sprinkler protection is required when the fire area is more than 12,000 square feet or the occupant load is 300 or more persons. (See Figure 9-9) In Group A-2 occupancies, the requirement for when an automatic sprinkler system is more restrictive when compared to other Group A occupancies because the fire area threshold is reduced to 5,000 square feet and the occupant load threshold is reduced from 300 to 100 or more. This lower occupant load is prescribed because alcoholic beverages are served as part of the business and the concern is the ability of impaired individuals to recognize a fire and rapidly

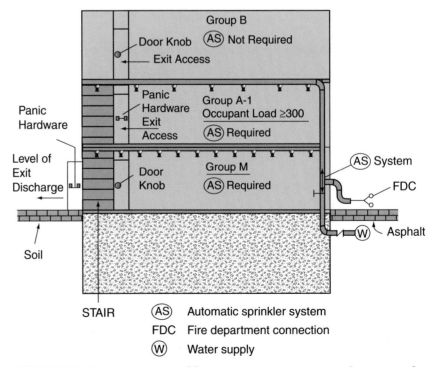

FIGURE 9-9 Automatic sprinkler system requirements for a mixed occupancy Group A-1/B/M occupancy

egress. In all of these occupancies, sprinkler protection is required in the fire area housing the occupancy. If the occupancy is located on a floor that is not the level of exit discharge, sprinkler protection must be extended to each floor between the Group A occupancy and the level of exit discharge. [Ref. 903.2.1.1, 903.2.1.2, 903.2.1.3, 903.2.1.4]

In Group B occupancies, automatic sprinkler protection is required only in buildings housing ambulatory health care facilities or in a building with this occupancy classification that has an occupant load of 30 or more that is located more than 55 feet above the lowest level of fire department vehicle access. (See Figure 9-10) The IFC exempts airport control towers, Group F-2 occupancies and open parking structures from these requirements. [Ref. 903.2.2 and 903.2.11.3]

While each occupancy classification has its own specific requirements based on specific hazards that may be located within them, Group F-1, M, and S-1 occupancies require automatic sprinkler protection when:

FIGURE 9-10 This B occupancy requires automatic sprinkler protection because it has an occupant load of more than 30 and the building is more than 55 feet above the lowest level of fire department vehicle access

1. the fire area of any one of these occupancies exceeds 12,000 square feet,
2. any of these occupancies is located more than three stories above the grade plane, or
3. the combined area of any one these occupancies on floors of a building, including mezzanines, exceeds 24,000 square feet.

These occupancies are treated as having equivalent fuel packages and loads unless a unique hazard is introduced within them. (See Figure 9-11) [Ref. 903.2.4, 903.2.7 and 903.2.9]

Buildings are assigned a Group H occupancy classification whenever they will store or use hazardous materials in excess of the limits permitted in IFC Chapters 27 through 44. Quantity limits, termed Maximum Allowable Quantity per Control Area (MAQ), are established in IFC Chapter 27. (Additional information about MAQs and how they are applied are in Chapter 16 of this text.) The Group H occupancy requirements in the IFC and IBC are not based on the occupant load or fire area—it is dependent on the hazard classification of the hazardous materials and the amount stored or used inside a building. When a building is constructed to the IBC Group H-1, H-2, H-3 or H-4 occupancy requirements, automatic sprinkler protection is required throughout the fire area. (See Figure 9-12) Automatic sprinkler protection

FIGURE 9-11 A Group S-1 occupancy requires automatic sprinkler protection when its fire area exceeds 12,000 square feet

FIGURE 9-12 A Group H-3 compressed gas packaging plant. All Group H occupancies require an automatic sprinkler system

is required throughout buildings classified as Group H-5. The design of the automatic sprinkler system must comply with the requirements in IFC Chapter 27 and for many hazardous materials its design will require special fire protection considerations. [Ref. 903.2.5]

FIRE DEPARTMENT CONNECTION

A fire department connection (FDC) is required for most NFPA 13 and 13R automatic sprinkler systems and standpipe systems. They are not required for automatic sprinkler systems protecting one- and two-family dwellings and townhomes. Fire apparatus can connect supply hoses to the FDC to pump additional water into sprinkler or standpipe systems. The piping arrangement between the FDC and the sprinkler riser depends on the type of automatic sprinkler system. [Ref. 903.3.7]

The location of and the fire hose threads installed on a FDC must be approved by the fire code official. The FDC's placement must not obstruct access to the protected building for other responding apparatus. The connection is located on the street side of buildings, and it must be easily recognized from the fire department vehicle access roadway. (See Figure 9-13) [Ref. 912.2]

To ensure that the sprinkler or standpipe system is supported by fire apparatus, it is important that its location be identified. For existing

FIGURE 9-13 Fire department connection for a wet-pipe automatic sprinkler system

FIGURE 9-14 FDC identification sign *(Courtesy of the City of Phoenix (AZ) Fire Department)*

buildings, the IFC authorizes the code official to require the installation of additional signs to help identify the FDC location. (See Figure 9-14) In many cases, the IFC and IBC only require automatic sprinkler protection for the fire area of an occupancy. In such cases, a sign is required at the FDC to indicate the portion of the building served. In Group R-2 apartment complexes with multiple buildings it is common to have a single FDC supplying multiple buildings, because it is more economical to provide separate water supply connections to serve a limited number of buildings, and in these cases, the buildings served by the FDC should be identified. **[Ref. 912.2.2 and 912.4]**

Fire Alarm and Detection Systems

Smoke alarms for occupant notification in one- and two-family dwellings and townhomes, local fire alarm and detection systems that notify building occupants, and automatic sprinkler systems connected to a supervising central station are examples of fire alarm and detection systems regulated by the IFC. Fire alarm and detection systems provide early warning of a fire by detecting fire's products of combustion such as smoke, heat, or a visible flame. Pre-action and deluge sprinkler systems and most alternative-agent fire extinguishing systems are activated by an automatic fire detection system connected to a fire alarm and detection unit. These systems are specified by the IFC in occupancies that present high life safety risks or buildings where a large population is present and require a reliable means of early occupant notification and communication.

DESIGN AND INSTALLATION STANDARDS

Requirements for the design, construction, and maintenance of fire alarm and detection systems are set forth in NFPA 72, *National Fire Alarm Code*. These systems utilize electricity, so the wiring of these systems must also comply with the NFPA 70, *National Electrical Code©*. NFPA 72 stipulates the requirements for components for initiating a fire alarm signal, signal transmission to and from the fire alarm control unit, occupant notification and audibility of an alarm signal, as well as performance, reliability, and survivability of the fire alarm and detection system. (See Figure 10-1)

An IFC compliant fire alarm and detection system is designed to perform several functions:

- Notification of a fire alarm in a building
- Monitoring and notification of supervisory and trouble signals
- Alerting the occupants
- Summoning aid
- Controlling of fire safety functions **[Ref. 902.1]**

In the context of NFPA 72, notification is an audible, visual or text signal, message, or display. Requirements for when notification devices are required are based on the building occupancy classification and occupant load. Audible alarm notification appliances must emit a distinctive signal that cannot be used for any other purpose other than that of a fire alarm. (See Figure 10-2) In any fire alarm and detection system, occupant notification takes precedence over any supervisory signals. **[Ref. 907.6.2.1]**

Supervisory signals indicate the need for action in connection of supervising guard tours, fire suppression systems, or maintenance. Supervisory signals include the closing of a main water supply control valve to a fire pump or standpipe system, low air pressure on a dry-pipe automatic sprinkler system, or any condition that takes a fire protection system off-line. A supervisory signal must be distinctive from other signals

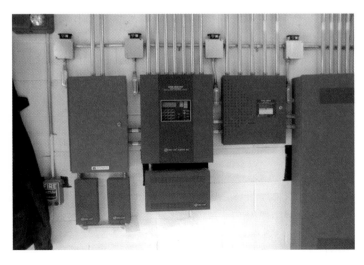

FIGURE 10-1 Fire alarm control unit

FIGURE 10-2 Audible and visual occupant notification device

FIGURE 10-3 Valve tamper switches that electrically supervise the wall post indicator valves for automatic sprinkler systems

that are received and must be visually annunciated at a fire alarm control unit. A supervisory signal warrants implementation of the building's fire protection impairment program. (See Figure 10-3) **[Ref. 902.1]**

Fire alarm and detection systems also can perform fire safety functions. A fire safety function is any feature that improves occupant life safety or controls the spread of fire and smoke. Fire safety functions include activation of motorized controls to close fire and smoke dampers or the release of magnetic devices that hold open opening protectives such as fire doors assemblies that protect openings in fire walls or barriers. (See Figure 10-4A) Activation of devices performing fire safety functions in buildings that are required to be equipped with a fire alarm and detection system must transmit a distinct audible and visual supervisory to a constantly attended location or activate the occupant notification devices. **[Ref. 907.4]**

FIGURE 10-4A Magnetic door hold-open device serving a fire safety function. Upon activation of the fire alarm and detection system, the magnet is de-energized, causing the fire door to close

FIGURE 10-4B Duct smoke detector *(Courtesy of Air Products and Controls, Pontiac, MI)*

FIGURE 10-5A A digital alarm communications transmitter that transmits fire alarm, supervisory, and trouble signals to a supervising station *(Courtesy of Honeywell Security and Communications)*

Duct smoke detectors perform fire safety functions. (See Figure 10-4B) These specialized smoke detectors are required by Section 606.2 of the *International Mechanical Code* when the air flow rate in a return air flow duct or plenum system is more than 2,000 cubic feet/minute. Duct smoke detectors constantly monitor the air for the presence of smoke in the return air duct or plenum. The detector is listed for use in mechanical ventilation systems based on the duct diameter or width, the air flow velocity, and the air temperature and humidity range of the air handling system. Activation of a duct smoke detector generally shuts down the air handling system, unless it serves a building smoke control system—in this case, the building's mechanical system is switched to a smoke control mode. **[Ref. 907.4.1]**

All IFC required fire alarm and detection systems are required to be monitored by an approved supervising station. (See Figure 10-5A) A supervising station is a facility that receives fire alarm and supervisory signals and transmits them to the fire department. The station is constantly staffed to respond to and process alarm signals. Monitoring is not required for smoke alarms or dwelling fire sprinkler systems in one- and two-family dwellings, nor is monitoring required for smoke detectors in Group I-3 occupancies. The design, installation, and maintenance of the monitoring circuits must comply with NFPA 72. **[Ref. 907.7.5]**

NFPA 72 requires certified and competent fire protection system designers and installers in the design, installation, testing, and maintenance of fire alarm and detection systems. The IFC requires the system be tested upon completion of the installation in accordance with NFPA 72. These test results are documented on a Record of Completion that is provided to the fire code official. **[Ref. 907.8]**

During the life of the occupancy, the owner is responsible for the continued maintenance of the fire alarm and detection system. Testing frequencies for the various types of detectors, the fire alarm control unit, audible and visual signal strength, and coverage and testing of devices performing fire safety are specified in NFPA 72. Within the first year of installation and every alternate year thereafter, all smoke detectors must be subjected to a sensitivity test. This test is performed to ensure smoke detectors are operating within their sensitivity range and listing. (See Figure 10-5B) **[Ref. 907.9]**

FUNDAMENTAL COMPONENTS

All fire alarm and detection systems required by the IFC have four fundamental components:

- Fire alarm control unit
- Initiating devices
- Occupant notification devices
- A primary and secondary electrical power supply

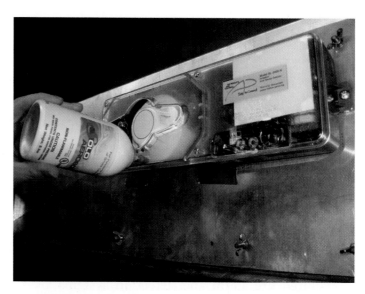

FIGURE 10-5B Functional test of a duct smoke detector (*Courtesy of Air Products and Controls, Pontiac, MI*)

The fundamental components are required for all protected premises, as defined in NFPA 72. These components are not found in single or multiple station smoke alarms required in Group R or certain I occupancies. All components of a fire alarm and detection system must be compatible with the fire alarm control unit, approved by the fire code official and listed by a nationally recognized testing laboratory. [Ref. 907.1.3]

A fire alarm control unit is the component that receives input from automatic and manual alarm initiating devices. The control unit may be capable of supplying electrical power to detection devices, notification appliances, and transponders that transmit signals to a supervising station. A fire alarm control unit can serve part or all of a building. For pre-action and deluge sprinkler systems and many alternative fire-extinguishing systems, the control unit must be listed as a releasing service fire alarm control unit, because it controls the activation of a fire extinguishing system. (See Figure 10-6) A fire alarm control unit must actuate notification devices and fire safety functions and annunciate the location of the initiating device within 10 seconds after its activation. The fire alarm control unit is used by firefighters to identify the location of the device that initiated the fire alarm signal. Unless a building is continuously occupied or is protected throughout by an automatic sprinkler system, the IFC requires that all fire alarm control units be protected by an approved automatic fire or smoke detection system. [Ref. 902.1 and 907.5.1]

Fire alarm control units are required by NFPA 72 to have a primary and secondary power supply. The primary power supply is commonly an electric utility. The source of the power is required to be supplied by a dedicated fire alarm branch circuit. The branch circuit must be protected from mechanical impact and identified at the circuit breaker so that power is not inadvertently turned off.

FIGURE 10-6 A fire alarm control unit listed for releasing service of alternative automatic fire-extinguishing system

FIGURE 10-7 A sprinkler water flow switch and valve tamper switch are examples of initiating devices. The control valve tamper switch is a supervisory device. Wiring for these devices terminates at the fire alarm control unit

The secondary power supply can be any source allowed by NFPA 72, and it is commonly accomplished using storage batteries. The secondary power supply must be capable of supplying power for at least 24 hours and, in the event of system activation, be able to operate all notification appliances for at least 5 minutes. If the fire alarm control unit is serving an emergency voice/communication system, NFPA 72 requires that the secondary power source be capable of operating for a minimum of 15 minutes. **[Ref. 907.7.2]**

Initiating devices are connected to the fire alarm control unit. Initiating devices are components that originate a change of state condition. Initiating devices include photoelectric or ionization smoke detectors, heat detectors, a manual fire alarm box, a sprinkler water flow or pressure switch, or a supervisory signal. (See Figure 10-7) NFPA 72 requires initiating devices be located so they are accessible for maintenance and testing and in all areas and locations prescribed by the IFC. If they can be subject to mechanical damage, NFPA 72 requires the initiating device be protected with a mechanical guard. **[Ref. 907.5]**

One initiating device that is accessible to all building occupants are manual fire alarm boxes. (See Figure 10-8) A manual fire alarm box is a device used by the public to initiate a fire alarm signal. Manual fire alarm boxes are normally required by the IFC in any occupancy where a fire alarm and detection system is required. If the building is protected by an automatic sprinkler system, manual fire alarm boxes are not required, provided the occupant notification devices activate when the automatic sprinkler system operates. **[Ref. 907.1]**

Manual fire alarm boxes must be located within 5 feet of an exit and the travel distance to each adjacent fire alarm box is limited to 200 feet. Boxes must be installed 42 to 48 inches above the finished floor and

FIGURE 10-8 Dual action manual fire alarm box

must be red in color. If the fire alarm and detection system is not monitored by a supervising station, a sign is required above each box stating that the fire department must be notified when the fire alarm box is activated. [**Ref. 907.5.2**]

Occupant notification appliances are required for many of the fire alarm and detection systems prescribed by the IFC. Notification appliances are installed in notification zones, which is an area of a building where all of the notification appliances simultaneously operate when an alarm signal is received and processed by the fire alarm control unit. Notification appliances are designed to deliver audible, visual, tactile, or a combination of these signals. (See Figure 10-9) [**Ref. 902.1**]

Occupant notification appliances activate when a initiating device operates. The IFC requires occupant notification appliances transmit an audible and visual signal. The audible alarm signal must be distinct and cannot be used for any other purpose. Audible notification appliances must be located so they can be heard above the sound level in the building. [**Ref. 907.6.2**]

FIGURE 10-9 Audible and visual occupant notification appliance

In Group A-1 occupancies with an occupant load of 1,000 or more and high-rise buildings require an emergency voice/alarm communication system. This occupant notification system is activated in the same manner as for audible and visual occupant notification appliances. However, this system requires the installation of a speaker network that delivers recorded or live verbal messages to the building occupants. This allows firefighters to direct occupants to perform either a complete, partial, or staged evacuation. In high-rise buildings, the emergency voice/alarm communication system must transmit signals to the floor where the alarm initiating device operated and the floor above and below this level. (See Figure 10-10) Individual paging zones are also required for each floor, exit stairways, at elevator groups and areas of refuge. The system must be designed so it can be manually controlled by emergency responders and be connected to an emergency power source. [**Ref. 907.6.2.2**]

OCCUPANCIES REQUIRING FIRE ALARM AND DETECTION SYSTEMS

The IFC sets forth requirements for fire alarm and detection systems in new and existing buildings. The IFC specifies the installation of automatic fire detection in occupancies where early detection and warning of an incipient fire is critical to occupant life safety. In certain existing buildings, the IFC requires the retroactive installation of these systems if they are not currently protected by a fire alarm and detection system. Requirements for when occupant notification depend on the occupancy classification and occupant load of the building. The IFC has specific requirements for these systems in certain buildings such as covered mall, underground, and high-rise buildings. They are not required in Group S and U occupancies. [**Ref. 907.2**]

FIGURE 10-10 An emergency voice/alarm communication system integrated into a fire alarm control unit
(Courtesy of Siemens Building Technologies, Florham Park, NJ)

The IFC allows the elimination of manual fire alarm boxes when the building is protected throughout by an automatic sprinkler system designed in accordance with NFPA 13 and the occupant notification appliances activate upon sprinkler water flow. However, for the purposes of testing, at least manual fire alarm box is required at an approved location. This manual fire alarm box is not required if the fire alarm system is for elevator recall service or in Group R-2 occupancies unless it is required by the fire code official. **[Ref. 907.2]**

Assembly occupancies require a manual fire alarm system that activates the occupant notification system when the occupant load is 300 or more. (See Figure 10-11) If the building has an occupant load of 1,000 or more, occupant notification must be performed using a emergency voice/alarm communication system. **[Ref. 907.2.1]**

With the exception of Group B occupancies housing an ambulatory health care facility, Group B and M occupancies only require a manual fire alarm system when the combined occupant load on all floors is 500 or more or if the occupant load is more 100 persons above or below the level of exit discharge. **[Ref. 907.2.2.1 and 907.2.7]**

All Educational occupancies with an occupant load of 50 or more require a manual fire alarm system that activates the occupant notification system. When these buildings are not protected by an automatic sprinkler system, manual fire alarm boxes can be eliminated when the interior corridors are protected by smoke detectors, auditoriums, cafeterias, gymnasiums, shops, and laboratories are protected by an approved means of fire detection, and the occupant notification system can be activated at a central point. (See Figure 10-12) **[Ref. 907.2.3]**

Group I occupancies represent a high life safety risk because patients may not be capable of self rescue. In the case of correctional

FIGURE 10-11 A Group A-3 occupancy with an occupant load of 300 or more requires a manual fire alarm system with a occupant notification system

FIGURE 10-12 Group E occupancy

and detention facilities, public safety mandates the housed individuals be closely supervised. Accordingly, the IFC requires the installation of automatic smoke detection that activates the occupant notification system. Activation of any other IFC prescribed fire protection systems also must activate the occupant notification system in these occupancies. [Ref. 907.2.6]

In certain critical health care areas, it may be desirable to reduce or completely eliminate notification appliance audibility. In intensive care or surgery suites, audible notification appliances can compromise patient care. In these cases, the design professional can use private mode signaling with the approval of the fire code official. In private mode signaling, the sound pressure level is reduced or may be eliminated. NFPA 72 requires the installation of visual notification appliances when private mode signaling is used. (See Figure 10-13) [Ref. 907.2.6, Exception 2]

Assisted living and board and care facilities are classified by the IBC as a Group I-1 occupancy. (See Figure 10-14) Because all Group I occupancies require the installation of a NFPA compliant automatic sprinkler system using listed quick response or residential sprinklers in patient sleeping areas, smoke detection is not required in habitable spaces, which are spaces in a building for living, sleeping, eating, or cooking. Bathrooms, closets, and storage or utility spaces are not habitable spaces. An automatic smoke detection system is required in corridors and waiting areas open to corridors. Single- and multiple station smoke alarms are required in the patient sleeping area, unless the area is served by the automatic smoke detection system. [Ref. 907.2.6.1]

Hospitals, intermediate care, skilled nursing homes, and detoxification facilities and other Group I-2 occupancies require automatic smoke detection systems. Smoke detection devices are required in corridors and spaces that are permitted by the IBC to be open to corridors, such as patient visiting areas and nurse stations.

FIGURE 10-13 Visual signaling device used as an element of private mode signaling
(Courtesy of Cooper Notification Inc.)

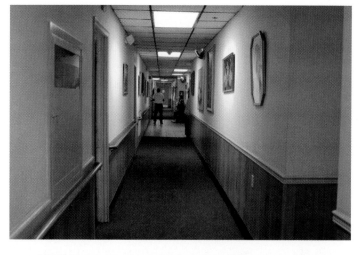

FIGURE 10-14 Group I-1 assisted living facility

FIGURE 10-15 Patient
sleeping room visual display
*(Courtesy of West Com Nursing
Call Systems, Fairfield, CA)*

FIGURE 10-16 Nurse call system with a graphic
user interface *(Courtesy of West Com Nursing Call Systems,
Fairfield, CA)*

In Group I-2 occupancies automatic smoke detection is not required in corridors of smoke compartments protecting patient sleeping rooms when each sleeping room is protected by a smoke detector. Activation of the sleeping room smoke detector in turn activates a visual display located outside of the patient's room in the corridor. The smoke detector's activation must initiate an audible and visual alarm at the nurse station that attends to the care of the patient. (See Figures 10-15 and 10-16) **[Ref. 907.2.6.2, Exception 1]**

A second option in Group I-2 occupancies is integrating the smoke detector into a patient sleeping room door closer. (See Figure 10-17) Activation of the smoke detector releases the door closer, causing the patient's room door to close and the occupant notification system. **[Ref. 907.2.6.2, Exception 2]**

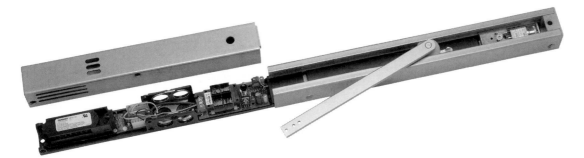

FIGURE 10-17 Door closer with an integral smoke detector *(Courtesy ASSA ABLOY Door Security Solutions, New Haven, CT)*

Group R occupancies regulated by the IBC require a fire alarm and detection system. (See Figure 10-18) However, the IFC requires an automatic sprinkler system that complies with either NFPA 13 or NFPA 13R in these occupancies, with occupant notification devices which activate upon sprinkler water flow. In Group R-1 and R-4 occupancies, at least one manual fire alarm box is required at an approved location. They are installed near the fire alarm control unit as a means of testing the fire alarm and detection system. In Group R-1 occupancies, an automatic smoke detection that activates the occupant notification system is required throughout all interior corridors that serve sleeping units. **[Ref. 907.2.8, 907.2.9, and 907.2.10]**

FIGURE 10-18 Group R-2 apartment building

Means of Egress

The IFC sets forth requirements for the fire protection and life safety of building occupants. The greatest concern and highest safety priority are the building occupants. The IBC specifies requirements for the design of a exit, exit access and level of exit discharge in all buildings. The IFC prescribes detailed regulations so individuals who are capable of rescuing themselves and other building occupants can safely and expeditiously leave a building in a fire or other emergency. The requirements found in IFC Chapter 10 are termed means of egress.

The IFC has requirements for means of egress in two chapters. The requirements in IFC and IBC Chapter 10 are the same and are applied to any new building or occupancy. A second set of requirements in IFC Chapter 46 is applied to existing buildings. The means of egress requirements in Chapter 46 are based on the occupancy classification and adopted building code at the time of the building's construction. In some jurisdictions, this may be an earlier edition of the IBC, one of the "legacy"[1] codes, or the community's own building code. If a conflict arises between the requirements in applicable building code and IFC Chapter 46, the most restrictive requirement is applied. In cases where a building is built in a jurisdiction that did not adopt a building code at the time of the construction, the means of egress system must comply with the requirements in Chapter 46 and be subjected to a life safety evaluation. This evaluation is submitted to the fire code official in the form of a technical report and opinion based on the requirement in Section 104.7.2. **[Ref. 4604.1]**

INTRODUCTION TO MEANS OF EGRESS

A building means of egress system has three distinctive and connected components:

- Exit access
- Exit
- Exit discharge **[Ref. 1002.1]**

Exit access is *that portion of a means of egress system that leads from any occupied portion of a building or structure to an exit.* (See Figure 11-1) Exit access is the area of room or space where egress commences. For the most part, any walking surface inside a building is a component in the means of egress system. The space inside a hotel room, factory, shopping mall, or high-rise building is designated by the IBC as either an exit access or exit. The IBC and IFC means of egress requirements ensure that any building uses or fixtures do not impede, limit, or obstruct one's ability to safely leave a building provided that the number of persons within a given space or area do not exceed code prescribed occupant load limits and travel distances to the level of exit discharge. Exit access includes corridors which physically guide a person to an exit. Exit access must comply with the general provisions for all means of egress components including ceiling height, protruding objects, continuity, and walking surfaces. It also must comply with the doorway, travel distance, and corridor requirements for exits and exit access. **[Ref. 1002.1 and 1003]**

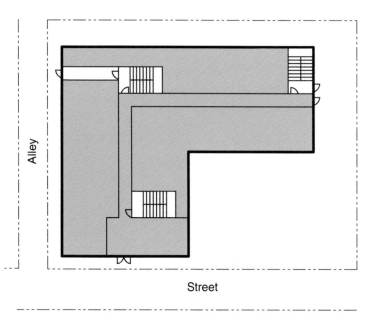

Shaded area = exit access

FIGURE 11-1 Exit access

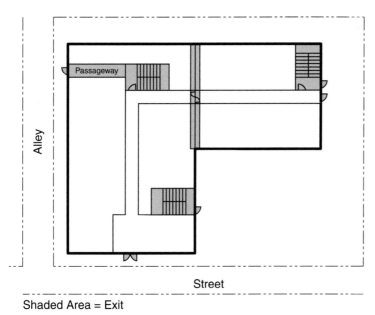

Shaded Area = Exit

FIGURE 11-2 Exit

The exit is *that portion of a means of egress system which is separated from other interior spaces of a building or structure by fire-resistance-rated construction and opening protectives as required to provide a protected path of egress travel between the exit access and the exit discharge.* (See Figure 11-2) An exit includes exterior exit doors at the level of discharge, vertical exit enclosures, exit passageways, exterior exit stairways and ramps, and horizontal exits. Exits cannot be used for any purpose that interferes with its function as a means of egress component. Once an exit is required to be constructed as a fire-resistive assembly or is equipped with automatic sprinkler protection, the required level of protection cannot be reduced until it terminates at the exit discharge. The minimum number of exits prescribed by the IBC is based on the occupant load per story. Depending on occupancy classification, only one exit may be required given the building occupant load and the location of the occupancy in relation to the building's grade plane. **[Ref. 1002.1 and 1021.1]**

A person has not completed egress from a building until they reach a public way such as a street or alley. Exit discharge is defined as *the portion of a means of egress system between the termination of an exit and a public way.* (See Figure 11-3) In a multi-story building, the level of exit discharge begins when the person reaches a grade level where the exit terminates and an exit discharge commences. An exit discharge must be at grade or provide a means of access grade. It cannot reenter the building. **[Ref. 1002.1]**

Because of the size of a building or the site's topography, it may not be possible to have a direct and unobstructed path to a public way. In this case, the IBC requires an area outside of the building that can safely

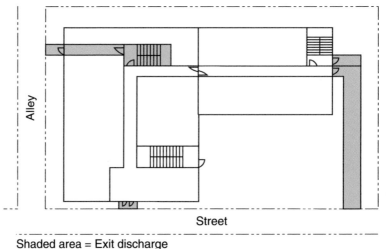

Shaded area = Exit discharge

FIGURE 11-3 Exit discharge

accommodate each occupant. A minimum area of 5 square feet is required for each person and the area must be located at least 50 feet from the building. The area must be permanently maintained and identified as a safe dispersal area. (See Figure 11-4A)

All IBC and IFC means of egress have minimum features and characteristics that must be properly designed and maintained compliant:

FIGURE 11-4A Safe dispersal area

1. Any given exit component requires a certain minimum width which is dictated by the occupant load and the IBC. [Ref. 1005.1]

2. A building's function is limited to a certain number of occupants in a given area. The number of persons allowed in the given area is the occupant load. Because buildings can have numerous functions, different occupant load factors are prescribed by the IBC and IFC. The total width of all exit access components must equal or exceed the occupant load. [Ref. 1004.1]

3. The IBC limits the distance from an exit to the exit access and the exit discharge. In some cases, the exit travel distances can be increased by providing fire-resistant construction in the exit access or by installing automatic sprinkler protection. All means of egress exit access must be spatially remote from one another and in most buildings, two or more means of egress are required. [Ref. 1016.1]

4. Exits, exit access, and the exit discharge are constructed on horizontal planes. Any change of elevation is accomplished using ramps, stairs, or steps. The IBC prescribes standard and consistent design criterion for these components to reduce and minimize tripping and fall hazards. [Ref. 1009.1 and 1010.1.]

5. The IBC requires the illumination and identification of exit access and level of exit discharge components. When the occupant load exceeds certain limits, emergency power for exit identification and illumination are required. [Ref. 1006.1 and 1011.1.]

6. To limit the potential for fire spread, the IBC prescribes more restrictive interior finish requirements to exit access components when compared to the requirements for rooms or enclosed spaces.

7. Generally, all means of egress require accessibility for mobility impaired individuals. Depending on the building's occupancy classification and level of fire protection, one or more areas of refuge may be required to shelter individuals who need rescue assistance. A means of communications is commonly required at the area of rescue assistance. (See Figure 11-4B) [Ref. 1007.1]

FIGURE 11-4B Sign indicating the location of rescue assistance for mobility impaired persons

FIGURE 11-5 An escalator and other types of people movers cannot be used as a component in a required means of egress system

8. A means of egress system that requires special knowledge or skills is not permitted by the IBC and IFC. A code compliant means of egress system simply requires persons to walk, operate simple door hardware, and follow one of the designated paths to the level of discharge. [Ref. 1008.1, 1014.1 and 1016.1]

9. Egress is complete when a person crosses the building property line and is in a public way. [Ref. 1002.1]

10. Unless used for accessible means of egress, people movers such as elevators, escalators, and moving sidewalks are prohibited from being components of a means of egress system. (See Figure 11-5) [Ref. 1003.7]

11. The building code official is charged with approving the design and construction of new or renovated means of egress including the occupant loads. The fire code official is responsible for ensuring the means of egress system is maintained.

OCCUPANT LOAD

Occupant load is the number of persons for which a means of egress is designed. With the exception of uses involving fixed seating commonly found in churches, stadiums, auditoriums, and restaurants, occupant loads are based on a density value which considers the number of square feet required for each person given a building's function or use. The occupant load is calculated by dividing the floor area being considered by the occupant per unit area factor value specified in IFC Table 1004.1.1 (See Table 11-1). Note that when applying this table that certain occupant load factors are based on the gross versus net floor area. Gross area is the area within the inside perimeter of exterior walls of building and excludes vent shafts, columns, and the thickness of interior walls. Net floor area is the actual occupied area and excludes unoccupied accessory uses including corridors, stairways, toilets, and mechanical rooms and closets. [Ref. 1004.1.1]

TABLE 11-1 Maximum floor area allowances per occupant (IFC Table 1004.1.1)

Function of space	Floor area in sq. ft. per occupant
Accessory storage areas, mechanical equipment room	300 gross
Agricultural building	300 gross
Aircraft hangars	500 gross
Airport terminal	
Baggage claim	20 gross
Baggage handling	300 gross
Concourse	100 gross
Waiting areas	15 gross
Assembly	
Gaming floors (keno, slots, etc.)	11 gross
Assembly with fixed seats	See Section 1004.7
Assembly without fixed seats	
Concentrated (chairs only—not fixed)	7 net
Standing space	5 net
Unconcentrated (tables and chairs)	15 net
Bowling centers, allow 5 persons for each lane including 15 feet of runway, and for additional areas	7 net
Business areas	100 gross
Courtrooms—other than fixed seating areas	40 net
Day care	35 net
Dormitories	50 gross
Educational	
Classroom area	20 net
Shops and other vocational room areas	50 net
Exercise rooms	50 gross
H-5 Fabrication and manufacturing areas	200 gross
Industrial areas	100 gross
Institutional areas	
Inpatient treatment areas	240 gross
Outpatient areas	100 gross
Sleeping areas	120 gross
Kitchens, commercial	200 gross
Library	
Reading rooms	50 net
Stack area	100 gross
Locker rooms	50 gross
Mercantile	
Areas on other floors	60 gross
Basement and grade floor areas	30 gross
Storage, stock, shipping areas	300 gross
Parking garages	200 gross
Residential	200 gross

TABLE 11-1 (Continued) Maximum floor area allowances per occupant (IFC Table 1004.1.1)

Function of space	Floor area in sq. ft. per occupant
Skating rinks, swimming pools	
Rink and pool	15 gross
Decks	50 gross
Stages and platforms	15 net
Warehouses	500 gross

In areas of buildings with fixed seating and aisles, the occupant load is calculated based on the number of seats. When the seats are not fixed, such as moveable tables and chairs, the occupant load is determined in accordance with the requirements in Section 1004.1.1. The calculated occupant load is added to number of fixed seats that may be available. Calculation of the occupant load with fixed seating will also depend on the seat's design—if it is without dividing arms, a load factor of 1.5 feet/person is used and for seating booths, an occupant load factor of 2 feet/person is applied. **[Ref. 1004.7]**

The occupant load must be posted in Group A occupancies. (See Figure 11-6) Posting the occupant load is essential for enforcement, especially when a fire department performs inspections of these occupancies during peak use. Many fire departments perform inspections of assembly occupancies at nights or weekends to ensure that the occupant load is not exceeded and that the egress system is functional. The occupant load sign must be conspicuously posted near the main exit or exit access door and be of a legible, permanent design. **[Ref. 1004.3]**

FIGURE 11-6 An occupant load sign

EGRESS WIDTH

Components of the means of egress must have an adequate width to safely accommodate the occupant load. Width is necessary to ensure the movement of the building occupants and reductions in width requires a reduction in the building occupant load. The width of means of egress components is specified by the IBC, such as the width of corridors or doors. In these cases, the width should be capable of safely accommodating the occupant load. To ensure that adequate capacity is built into the egress system, the IBC requires the calculation of egress width. The factors used in the calculation of exit width are dependent on the function of the component. For changes in grade or elevation that require the use of stairs, a factor of 0.2 inches/occupant is used. For all other egress components such as corridors, doors, and ramps, a factor of 0.3 inches/occupant is used. (See Figure 11-7) In multiple story buildings, the maximum capacity required from any story of the building must be maintained to the termination of the means of egress. (See Figure 11-8) **[Ref. 1005.1]**

Code Basics

Occupant load is based on the function or purpose of room or area. Depending on the function, it is calculated using either the net or gross floor area. In assembly occupancies, the maximum occupant load must be conspicuously posted. ●

FIGURE 11-7 The minimum width of an exit stair is 0.2 inches/occupant, while the exit width of an exit ramp exit is calculated using a factor of 0.3 inches/occupant

Given: A portion of the egress system serves 200 occupants in a unsprinklered Group B building.

Determine: The minimum required widths of each egress component.

Solution: Minimum Required Calculated Stair Width: (200) (0.3 inches/person) = 60 inches
Minimum Required Calculated Width of All Other Components: (200) (0.2 inches/person) = 40 inches

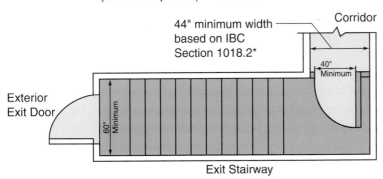

FIGURE 11-8 The required capacity of a means of egress system shall not be reduced by the egress path

A properly designed means of egress system is commonly built with two separate and remote paths. The IBC requirements for egress width follow this philosophy in that adequate width must be provided so that in the event one egress path is lost or compromised, the available width is reduced no more than 50% of the required capacity. [**Ref. 1005.1**]

If improperly installed, an egress door can create reduce the exit width. (See Figure 11-9) The IFC and IBC establish requirements for doors that encroach into the egress width to ensure the egress width is not reduced. When fully open,

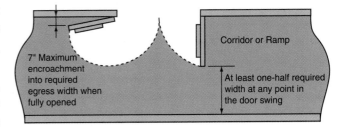

FIGURE 11-9 Door encroachment into an egress path

doors cannot reduce the width of an egress component by more than 7 inches. In any other position, the door cannot reduce the required width by more than 50%. Surface mounted latch release hardware, such as door knobs or release latches, are exempt from the 7-inch projection requirement if they face the corridor when the door is fully open. [**Ref. 1005.2 and 1005.3**]

EXIT ACCESS AND EXIT ACCESS TRAVEL DISTANCE

Another important component is the distance one must travel to safely egress a building. Maximum egress travel distances in the exit discharge path are established in the IBC and IFC. The exit access pathway must be clear and distinct and should avoid travel through intervening spaces that could confuse occupants attempting to egress a building. Travel distances and the geometry of exits and exit access are closely regulated in the IBC and IFC.

The intent of the IBC exit access requirements is exits should be direct from the room or area under consideration. However, this is not always possible, so the code grants exceptions for a limited number of circumstances where exits can pass through adjoining rooms or spaces rather than directly into corridors or exit enclosures.

Egress through an adjacent area or room is allowed provided the exit path is direct and obvious so that the occupant is cognizant of the exit travel path. When egress is through an intervening space, it must be under the same control as the space where egress commenced. When egress is through intervening spaces, the space under consideration must be accessory, meaning that its use is complimentary to use of the room or room where egress starts. The code is more concerned with the uses of the spaces rather than their area when evaluating this component of the means of egress. [**Ref. 1014.2**]

The exit access path cannot pass through a room or area that can locked. In addition, certain building uses present a very high probability of being obstructed, including closets, storage rooms, and kitchens. Accordingly, the IBC does not permit the use of these areas as exit access path. An exception is granted to Group M occupancies by allowing egress through stockrooms, because this occupancy consistently has a very low fire loss history and the occupants are cognizant and aware of their surroundings. (See Figure 11-10) However, the IBC sets specific limits, including that stored goods and commodities must have the same hazard classification as found in the main retail area, the egress path is demarcated by partial or full-height walls with a minimum width of 44 inches that leads to an exit, and no more than 50% of the exit access is through the stock room area. In addition, doors into the stockroom cannot be locked. [**Ref. 1014.2**]

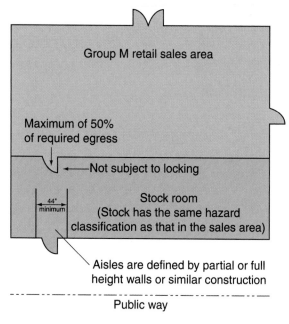

FIGURE 11-10 Exit access through a Group M occupancy stockroom

TABLE 11-2 Exit access travel distance (IFC Table 1016.1)[a]

Occupancy	Without sprinkler system (feet)	With sprinkler system (feet)
A, E, F-1, M, R, S-1	200	250[b]
I-1	Not Permitted	250[c]
B	200	300[c]
F-2, S-2, U	300	400[c]
H-1	Not Permitted	75[c]
H-2	Not Permitted	100[c]
H-3	Not Permitted	150[c]
H-4	Not Permitted	175[c]
H-5	Not Permitted	200[c]
I-2, I-3, I-4	Not Permitted	200[c]

(Footnotes not shown. See IFC Table 1016.1 for all footnotes)

The distance occupants must travel along the exit access is regulated by the IFC, and the requirements are based on the occupancy classification or a specific building's use, such as covered malls or atriums, and whether the building is equipped with an automatic sprinkler system. The requirements for travel distance are measured from any occupied point in the building to the nearest exit component rather than to all required exits from a room, floor, or building. Exit access travel distance is measured to:

- An exit passageway
- A horizontal exit
- The door of a vertical exit enclosure
- An exterior exit ramp or egress stairway
- An exterior exit door when it is located at the level of exit discharge (See Table 11-2) **[Ref. 1016.1]**

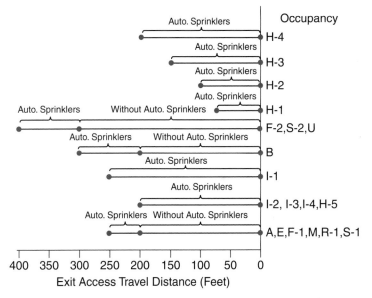

FIGURE 11-11 The permissible exit access travel distance can be increased in certain occupancies when either a NFPA 13 or 13R automatic sprinkler system is installed

Travel distance is measured along the logical and unobstructed path available to the occupants. (See Figure 11-11) The path will be modified by furnishings such as work stations, chairs, and the like, machinery, or fixtures such as shelves or storage racks. Some of these items may be moveable, which can complicate plan review or inspections. One conservative method that is reasonable in application is measuring using right angles, because it recognizes that obstructions such as machinery, racks, and desks will be circumnavigated by occupants. This method of measurement accounts for any future obstructions that may be located in the exit access travel path. **[Ref. 1016.1]**

Except in occupancies with small areas, a properly designed egress system commonly has two or more exits or exit access doorways. The

TABLE 11-3 Spaces with one exit or exit access doorway (IFC Table 1016.1)

Occupancy	Maximum Occupant Load
A, B, Eª, F, M, U	49
H-1, H-2, H-3	3
H-4, H-5, I-1, I-3, I-4, R	10
S	29

a. Day care maximum occupant load is 10.

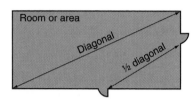

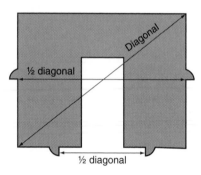

FIGURE 11-12 Example exit and exit access doorway arrangements

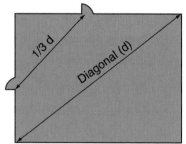

Building is protected by an automatic sprinkler system complying with IFC Section 903.3.1.1 or 903.3.1.2

FIGURE 11-13 The separation between exit and exit access doorways in sprinklered buildings can be reduced to 33.3%

IBC requires two such doorways when the occupant load exceeds the values prescribed in Table 1015.1 (See Table 11-3), if the common path of egress travel exceeds the distance limitations in IBC Section 1014.3, in rooms of a given area housing boilers, incinerators, furnaces or refrigeration machinery, and stages. As the building or space occupant load increases, the IBC prescribes a greater number of exit or exit access doorways. **[Ref. 1015.1]**

The IBC and IFC prescribe specific requirements relative to the location and arrangement of the required exit or exit access doorways in relation to one another. These requirements ensure with a high degree of reliability that if one means of egress is obstructed, the others will remain available and will be usable by the occupants. This approach assumes that since the remaining means of egress are still available, there will be sufficient time for the building occupants to use them to evacuate the building or space.

If two exits or exit access doorways are required, they must be arranged and placed a distance apart equal to not less than one-half of the maximum overall diagonal of the space, room, or building area served. The minimum distance between the two means of egress, measured in a straight line, shall not be less than one-half of that maximum overall diagonal dimension. (See Figure 11-12) **[Ref. 1015.2.1]**

In buildings protected throughout by an automatic sprinkler system complying with NFPA 13 or NFPA 13R, the separation of the exit or exit access doorways can be reduced. (See Figure 11-13) Because the building is protected by an automatic sprinkler system, the IBC and IFC allow distance between exits and exit access doorways to be reduced to 33 1/3% instead of the conventional 50% value prescribed for all building not protected by these systems. **[Ref. 1015.2.1]**

EXIT SIGNS AND MEANS OF EGRESS ILLUMINATION

An unwanted fire, natural, or technological disaster can impact the life safety of building occupants. One concern is the ability to visually locate the means of egress and clearly identify openings that allow occupants to quickly and easily pass through the level of exit discharge. The IBC requires illumination of the means of egress when the building is occupied and identification of exit and exit access doorways using signs constructed to specific standards that can be easily recognized and comprehended by the general population. With the exception of Group U occupancies, aisles in Group A occupancies, dwelling and sleeping units in Group R-1, R-2, R-3, and I occupancies, means of egress illumination is required for all occupancies. Illumination must be provided throughout the exit access, exit, and at the exit discharge. [Ref. 1006.1]

Exit illumination must emit enough visible optic energy to equal 1 footcandle at the walking surface. 1 footcandle of luminance equals 10.76 lux, which is the SI derived unit of measurement. Lux is commonly expressed as lumens, which is the amount of light the source produces and approximately equals 0.1 footcandle. 1 lux equals one lumen per square meter of area and 1 footcandle equals 1 lumen illuminating one square foot. 11 lux (or lumens) equal 1 footcandle. In terms of area, which is easier to quantify, 1 footcandle equals 0.30 footcandle^2. A 40-watt light bulb emits about 455 lumens or about 41.4 footcandle. [Ref. 1006.2]

While the probability of utility electric power failure is low, its failure is a credible event and the IBC and IFC specify requirements to ensure an alternative electrical power source is available for means of egress and exit sign illumination. (See Figure 11-14) Emergency power, which must operate within 10 seconds of power loss, is required for building egress components that require two or more exit or exit access doorways. This includes aisles, corridors, exit enclosures, and exit passageways, as well as interior and exterior exit discharge components. [Ref. 1006.3]

The source of emergency power must be capable of operating for a minimum of 90 minutes. The source of power can be storage batteries, unit equipment which is equipped with its own standby power source, or an engine-driven generator and it must be maintained in accordance with IFC Section 604. [Ref. 1011.5.3] (See Chapter 6 of this text for a review of the IFC, IBC, and NFPA requirements for standby and emergency power systems)

To clearly demarcate exit and exit access doorways, the IBC and IFC prescribe the installation of approved exit signs. Signs are required to indicate the direction for egress travel when the exit or exit access is not clearly or immediately visible to building occupants. Signs must be placed in accordance with their listed viewing distance but not more than 100 feet apart or in distance to the exit or exit access doorway. (See Figure 11-15) Signs are not required in

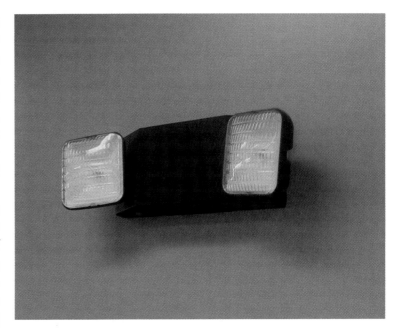

FIGURE 11-14 Emergency lighting unit

FIGURE 11-15 This listed exit sign includes a marking to indicate its maximum viewing distance *(Courtesy of ASSA ABLOY Door Security Solutions, New Haven, CT)*

FIGURE 11-16 UL 924 listed electrically illuminated exit sign

rooms or areas with only one exit or exit access door and, when approved by the building code official, in buildings with glass store fronts where the main exterior exit doors are clearly obvious and identifiable. Exit signs are not required in Group U occupancies and in sleeping or dwelling units of Group R-1, R-2, R-3, and I-3 occupancies. [Ref. 1011.1]

Exit signs are internally or externally illuminated. Internally illuminated signs can use electric lamps or with chemicals that are self-illuminating or are photo luminescent. All exit signs must be listed and labeled as complying with UL 924, *Standard for Safety of Emergency Lighting and Power Equipment.* (See Figure 11-16) Internally illuminated signs must remain illuminated at all times. In the event of a power loss, UL listed exit signs constructed as unit equipment are required to operate for at least 90 minutes at no less than 60% of its rated voltage. [Ref. 1011.4]

Externally illuminated exit signs are allowed by the IBC. When externally illuminated signs are used, they must meet the dimensional sizing requirements and be of a contrasting color. At the face of the sign, there must be a minimum luminance level of not less than 5 footcandles. The source of lighting must be connected to an approved emergency power source that can provide electricity for at least 90 minutes. [Ref. 1011.5]

FIGURE 11-17 The IFC recognizes the common practice of securing means of egress in unoccupied area of buildings. *(Courtesy of the Minnesota State Fire Marshal, St. Paul, MN)*

MEANS OF EGRESS MAINTENANCE

IFC Section 1030 establishes requirements for the maintenance of the means of egress system. These provisions address obstructing, blocking, or disabling means of egress components. The requirements also are intended to prevent conditions that can cause confusion or obscure the means of egress.

When a building is occupied, regardless of occupant load, the required exit components are required to be maintained free of any obstructions or impediments that can prevent the complete and instantaneous use of the means of egress. These requirements do not prevent the use of security devices that can render the means of egress unusable provided the building is not occupied. (See Figure 11-17) [Ref. 1030.2]

All components of the means of egress must be available in the event of an emergency that requires occupants to leave a building. The IFC prohibits any obstructions to exits, exit access, or the exit discharge. Such obstructions must be immediately removed. (See Figure 11-18A and B) [Ref. 1030.3]

FIGURE 11-18A and 11-18B Obstructions to means of egress components, such as shopping carts blocking an exit discharge opening or a chain and padlock on a gate in the path of exit discharge, are serious fire code violations

The IFC requires exit signs be installed in accordance with the Chapter 10 requirements. Signs in buildings must not be obstructed or blocked by drapes, decorations, or partitions, nor should there be other signs that can distract or prevent attention. (See Figure 11-19) [Ref. 1030.4]

FIGURE 11-19 Objects or signs that confuse or impair visibility or identification of an exit are prohibited

EXERCISE

Loretta's is a Group A-2 night club that serves alcoholic beverages. The business is moving into an existing wood frame building previously used as a convenience store. The building will be protected throughout by an automatic sprinkler system in accordance with IFC Section 903.3.1.1 and be renovated by the addition of fixed seating and a dance floor. (See Figure 11-20)

A permit application was submitted to the Fire Department for a Group A occupancy operating permit. Calculate the occupant load based on the amount of open space in the occupancy, the fixed seating occupant load, and the exit width based on the occupant load.

Loretta's Honky Tonk

Calculation Notes:
· Calculate the occupant load based on the number of fixed seats and the area of the dance floor.
· The occupant load factor for the dance floor is 7 sq. ft. / person.
· The egress width factor for the exit door is 0.2 inches / person
· The serving area behind the bar is 51 square feet and the exit aisle from the dance floor is 3 feet wide by 28 feet length.

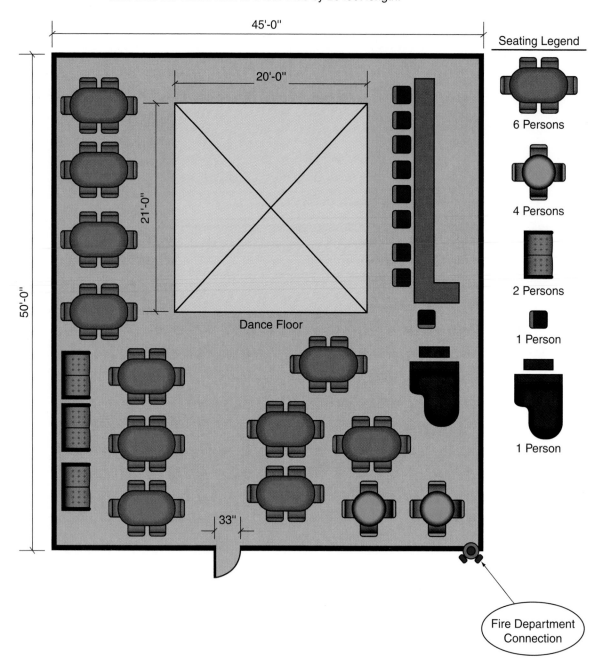

FIGURE 11-20 Means of egress exercise

SOLUTION
Occupant Load Calculation

Fixed Seats	Number	Occupants
6 person	11	66
4 person	2	8
2 person	3	6
1 person	10	10
TOTAL FIXED SEATS		90

Dance floor occupant load: $(420 \text{ Ft.}^2)/(7 \text{ Ft.}^2/\text{Person}) = 60$ persons
Bar service area occupant load: $(51 \text{ Ft.}^2)/(200 \text{ Ft.}^2/\text{Person})^* = 1$ Person
Exit aisle occupant load: $(168 \text{ Ft.}^2)/(5 \text{ Ft.}^2/\text{Person})^{**} = 34$ persons
Total occupant load: $90 + 60 + 1 + 34 = 185$ persons
* Based on IFC Table 1004.1.1 value for commercial kitchens
** Based on IFC Table 1004.1.1 value for assembly – standing area

Exit width calculation
$(33 \text{ inches})/(0.3 \text{ inches/person}) = 110$ occupants

Findings
1. The exit width cannot accommodate the occupant load. [Ref. 1005.1]
2. Based on the occupant load, a second exit or exit access doorway is required. [Table 1015.1]
3. Panic hardware is required for the exit doors. [Ref. 1008.1.10]

PART

V

Special Processes and Building Uses

Flammable Finishes

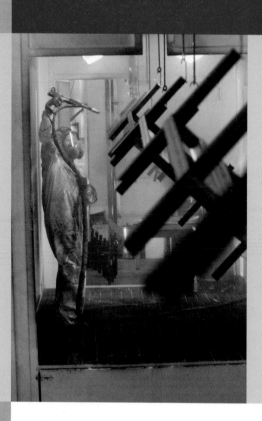

Unique processes and materials are used in the application of flammable finishes. Liquids can be atomized into an aerosol and sprayed onto objects or goods can be immersed into open tanks. In a process known as powder coating, ionized solids are suspended in air and attach to metal objects after being electrically charged. Rigid thermoplastics are manufactured using glass fibers that are applied using a resin and chemically activating it using a chemical catalyst. IFC Chapter 15 contains requirements that address processes where flammable and combustible liquids or combustible dusts are applied onto metal, wood, or plastic goods to provide aesthetically pleasing and durable coatings.

Section 1503 contains the minimum requirements for the protection of operations involved in the application of flammable finishes. This section addresses sources of ignition, storage, use, and handling of flammable and combustible liquids and the operation and maintenance of flammable finishing activities. Section 1503 is applicable to all of the activities regulated in Sections 1504 through 1510.

TYPES OF FLAMMABLE FINISHING PROCESSES

Paints and coatings applied by spray finishing can be hydrocarbon solvents or water based. They are formulated with film forming chemicals and pigments. Properly applied film forming chemical(s) bond together and create a seamless surface. The pigment is the color in the paint or coating. Generally, a solvent (either hydrocarbon or water) is added to the coating to carry the film formers and pigments to the surface onto which they are being applied, to initiate the polymerization of the film former and to accelerate drying. In many industries, solvent based paints and coatings are preferred for finish quality reasons and they generally dry quicker when compared to water-based paints. Hydrocarbon solvent products may release volatile organic compounds (VOCs) that are a fire and environmental hazard.

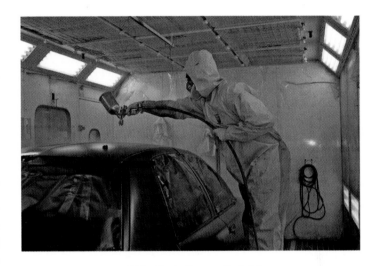

FIGURE 12-1 Spray finishing

Spray finishing regulated by the fire code is a process where a flammable or combustible liquid is pressurized and discharged through a very small orifice to create an aerosol. (See Figure 12-1) Aerosols are very small droplets of flammable or combustible liquids suspended in air. The droplets have very little mass but a large surface area. The droplets in air are easily ignited because they quickly evaporate into vapor. By discharging a large number of very small flammable or combustible liquid droplets under pressure, spray finishing produces a large volume of flammable vapor.

Powder coating is the application of thermoforming or thermosetting plastic powders onto a metal surface. (See Figure 12-2) Common powders include polyester, polyester-epoxy, or

FIGURE 12-2 Powder coating

acrylics. The powders range in size from 20 to 40 micron, which is comparable to the size of baking flour. Powder coating does not require a solvent to bond the powder onto the metal- it produces less pollution and essentially no hazardous waste when compared to petroleum solvent-based liquids.

After the surface is cleaned the powder is applied onto the metal surface, the powder receives a positive electrical charge from the powder coating gun or nozzle. The charged particles adhere to metal surfaces, which are negatively charged. The oppositely charged powder is attracted and adheres to the metal parts. Powder coating guns generally are used for manual application while powder coating nozzles are found on automatic or robotic equipment.

Suspension of the powder in air represents the greatest hazard. The powder is a combustible dust and can burn or contribute to a deflagration. A deflagration can occur when the concentration of the powder is within its minimum explosive concentration (MEC) and is suspended in air, a strong enough source of ignition is present and the dust is confined inside of a powder coating booth or room. The potential for dust deflagration extends to dust collection equipment, especially powder coating operations using remote dust collectors that are connected to the powder coating equipment using ducts.

Manufacturing of reinforced plastics involves the use of a glass-fiber mat (also known as "rove"), which is placed onto a mold of the desired shape. The glass-fiber mat is coated or sprayed with resin that is mixed with catalyst to accelerate the drying rate of the resin. Reinforced plastics are found in the construction of boats, plumbing fixtures such as sinks and bathtubs, and many automotive applications.

The most common resin that is used is unsaturated polyester resin (UPR). The primary constituent of UPR is styrene monomer. Most UPRs are classified as either Class IC flammable or Class II combustible liquids and may also be classified as either a Class 2 or Class 1 unstable (reactive) liquid.

A catalyst is added to promote drying and the uniform hardening of the resin. The catalyst most commonly used is the organic peroxide methyl ethyl ketone peroxide. (See Figure 12-3) Organic peroxides are a unique class of hazardous materials because they contain hydrocarbon molecules connected or branched to oxygen molecules. Organic peroxides are sensitive to any contamination and also have a limited shelf life based on the storage temperature.

During the manufacturing process of reinforced plastics, the resin and the catalyst will generally be mixed together manually or at the spraying or "chopper" gun. (See Figure 12-4) A chopper gun is a pneumatically-powered, hand-held, or robotically-controlled device that simultaneously applies glass fiber strands, resin, and the catalyst. The process can involve the manual smoothing of the resin, fibers, and glass fibers using paint rollers.

FIGURE 12-3 Methyl ethyl ketone peroxide is an organic peroxide that can be easily ignited and has a limited shelf life

FIGURE 12-4 Chopper gun used in the manufacturing of reinforced plastics

SPRAY BOOTH AND SPRAY ROOM CONSTRUCTION

IFC Section 1504 specifies requirements for the construction of spray spaces, booths, and rooms and the application of flammable spray finishes. IFC Section 1504 requirements stipulate the location of spray finishing operations, the design and construction of spray rooms, spray spaces and spray booths, fire protection, control of ignition sources, mechanical ventilation systems, interlocks required for spray finishing, and limited spray spaces. Section 1504.1 requires that spray finishing activities also comply with the requirements in Section 1503 for the protection of operations.

Spray finishing is allowed in a spraying space, spray booth, or spray room. In Group A, E, I, and R occupancies, spray finishing can only occur in a spray room that is separated by fire-resistive construction in accordance with the IBC and is protected by an automatic sprinkler system designed in accordance with Section 903.3.1.1. In all other occupancies, spray finishing can be performed in a spray booth. [Ref. 1504.2]

Regardless of the occupancy classification of the building where spraying is occurring, the IFC requires that all spraying rooms, booths, and spaces be:

- Constructed of approved materials
- Equipped with an approved mechanical ventilation system
- Properly illuminated using approved luminaires
- Equipped with required interlocks
- Designed to control sources of ignition
- Protected with an automatic fire extinguishing system
- Maintained as it relates to housekeeping and storage of hazardous materials

In some occupancies, the code official may approve *limited spraying spaces*. Limited spraying spaces commonly are found in automobile body shops where only small jobs are performed. In order to qualify as a limited spraying space:

- The aggregate surface area to be sprayed shall not exceed 9 square feet. This is large enough to allow the painting of hoods or trunk lids.
- Spraying operations can not be continuous.
- Positive mechanical ventilation providing a minimum of six complete air changes per hour is required and must meet the *International Mechanical Code* (IMC) requirements for hazardous exhaust systems.
- Electrical wiring within 10 feet of the floor and 20 feet horizontally of the limited spraying space shall be designed for Class I, Division 2 locations in accordance with the *National Electrical Code©*. [Ref. 1504.9]

FIGURE 12-5 A cross draft spray booth

A spray booth is an appliance that can be open on one or more sides to facilitate material handling or the movement of large parts. (See Figure 12-5) It is important to remember that a spray booth is an appliance and not an occupancy- the IBC Group H occupancy requirements do not apply. [Ref. 1502.1]

All spray booths requires mechanical ventilation. A properly designed and maintained mechanical ventilation system provides sufficient supply and exhaust air to maintain the atmosphere inside of a spray booth or area below 25% of the lower flammable limit of the most volatile flammable liquids sprayed. The mechanical ventilation system dilutes the flammable vapor with air by mixing it inside of a plenum and safely discharging the vapor/air mixture outdoors or into a pollution control device. Dry or wet filters are provided to capture pigments and film formers and prevent them from being conveyed through the mechanical ventilation system.

Spray rooms, booths, and spraying spaces are constructed of approved noncombustible materials. The IFC requires the interior wall surfaces and any surfaces where accumulations can be deposited to be smooth in construction so as to allow free air movement and to facilitate cleaning. The IFC prohibits the use of aluminum in the construction of spraying rooms, booths, or areas. Aluminum exhibits a much lower melting point when compared to carbon steel—aluminum would limit a booth's ability to confine a fire. [Ref. 1504.3.2.2]

Walls, floors, and ceilings, as well as their exhaust ducts and all associated components that form the spray booth appliance, are required to be constructed of noncombustible materials. In most cases, minimum 18 gage sheet metal is used. However, some manufacturers will create a structural panel by creating a wall assembly using structural steel channel that is covered by 20 gage steel. Spray booths constructed using sheet metal wall and ceiling components are shipped unassembled and are erected on site. To ensure that the air flow remains uniform during the operation of spray booths, the IFC allows wall and ceiling joints to be sealed by caulks or other similar sealants. (See Figure 12-6) [Ref. 1504.3.2.2]

FIGURE 12-6 A downdraft spray booth constructed of approved, noncombustible materials

MECHANICAL VENTILATION

Of all of the safety features installed in a spray booth or room, mechanical ventilation has the greatest positive impact on the protection of the booth, its occupants (when occupied), and the goods being finished. A properly designed and maintained mechanical ventilation system safely removes flammable vapors or combustible dusts and exhausts them to a safe location. The mechanical ventilation system maintains the atmosphere inside of the flammable vapor area below 25% of the lower flammable limit for the most volatile flammable liquid or 50% of the minimum explosive concentration for combustible dusts.

To minimize the fire or deflagration risk, mechanical ventilation systems for spray finishing areas are generally designed to exhaust at a rate of 10,000 cubic feet/minute for each gallon of liquid spilled. (See Figure 12-7) At 10,000 CFM, this provides a four-to-one (4:1) safety factor in the design of the mechanical ventilation system. NFPA 33, *Code for Spray Application Using Flammable or Combustible Materials* requires the design of a mechanical ventilation system for a spraying area maintain the atmosphere at or below 25% of the lower flammable limit for the most volatile liquid being sprayed. The design of these systems must also comply with the hazardous exhaust system requirements in IMC Section 510. [Ref. 1504.7]

When compared to other mechanical ventilation systems, hazardous exhaust systems have unique design requirements:

- The ventilation system design must use uncontaminated supply air to dilute and maintain the flammable constituents at concentrations below 25% of its lower flammable limit.
- The exhaust system, including the exhaust duct, fans, and fan motors, must be independent of other building mechanical exhaust systems. With exceptions for laboratories, hazardous exhaust systems shall not share common shafts with other air-handling duct.

FIGURE 12-7 Mechanical ventilation systems for spray booths and spray rooms are designed to maintain the atmosphere at less than 25% of the lower flammable limit and in accordance with the IFC and IMC

- A hazardous exhaust system for a flammable vapor area must be designed using the constant velocity method. The IFC specifies a minimum air flow capture velocity of 100 feet/minute based on the cross-sectional area of the spray room or booth. **[Ref. 1504.7.3]**
- A condition of acceptance is air flow balance testing. In flammable vapor areas, balancing is accomplished by ensuring the volume of makeup air is approximately equal to the rate that air is exhausted. One economical method of verifying this is requiring an air balance test report. An air balance test report documents the measurements of the air velocity in the supply air source and the exhaust duct to ensure the exhaust fan is discharging at its design capacity. For example, if a spraying area is required to be protected by an exhaust fan rated at 15,000 cubic feet minute at a given static air pressure, the air balance test report will document if the fan is performing as specified. Air balance test reports can be prepared by a booth manufacturers' representative or heating, ventilating, and air conditioning firms. (See Figure 12-8)

FIGURE 12-8 An air balance test can accurately and economically confirm the hazardous exhaust system complies with the IFC and IMC

Because the mechanical ventilation system for a spraying area must comply with IMC Section 510, independent exhaust ducts are required from each spray booth, room, or area. (See Figure 12-9) The IFC requires the exhaust duct meet the certain separation distances from the point of discharge:

- Property lines: 30 feet
- Openings into buildings: 10 feet
- Height above exterior walls and roofs: 6 feet
- Combustible walls or openings into the building that are in the direction of the exhaust discharge: 30 feet
- Above adjoining grade: 10 feet **[Ref. 1504.7.6]**

To ensure the velocity of the exhaust air meets the 100 linear foot/minute flow rate specified in Section 1504.7.3, the IFC requires a means of measuring air flow be provided using a visual gauge, pressure switch, or an audible alarm to indicate a reduction in the required air flow velocity. The most common method used is the installation of a manometer that measures the air pressure at the intake and exhaust side of the booth's exhaust plenum. The manometer measures the differential pressure of the exhaust air. A common unit of measurement is inches of water column. (27.71 inches of water equals one pound/square inch gauge).

As the filter becomes loaded with film forming chemicals and pigments, the differential pressure increases. The manometer measuring

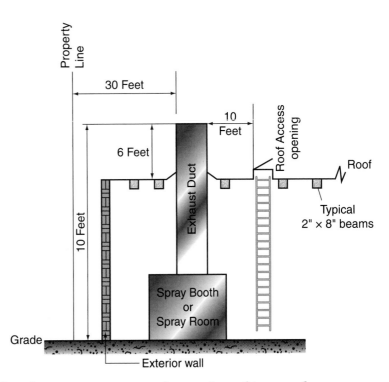

FIGURE 12-9 The exhaust system for spray booth or spray room must be terminated in accordance with the IFC and IMC requirements

tube on the intake side (the spraying area) will register the increased pressure that results from the increased resistance to air flow caused by the filter debris. When the differential pressure exceeds the manufacturer's pressure limits, the filters should be replaced. (See Figure 12-10) **[Ref. 1504.7.8.3]**

ILLUMINATION

To facilitate the application of flammable finishes, the interior of spraying spaces are illuminated by electrical lamps. The use of electric lamps creates a potential ignition source because of the heat they produce or when lamps are damaged, their energized circuits can be exposed. The IFC prohibits portable electric lamps for the illumination of flammable vapor areas and requires that when portable lamps are used, they must be limited to cleaning or repair operations and approved for hazardous (classified) electrical atmospheres. **[Ref. 1504.6.2.4]**

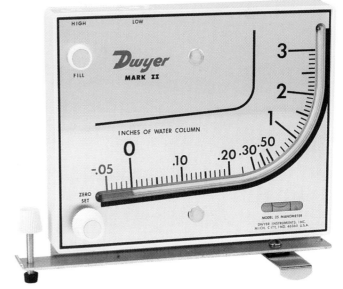

FIGURE 12-10 Differential Pressure Manometer
(Courtesy of Dwyer Instruments Inc., Michigan City, IN)

Fixed lamps or luminaires illuminate the interior of a spraying area. The requirement for lamps depends on their design. Luminaires are either external or internal (termed as "integral" in the IFC). External luminaires are designed where the light bulbs are in an area that is not classified as a hazardous location. Internal (integral) luminaires are designed where the light bulbs are located and replaced in an area within a hazardous (classified) location. Hazardous (classified) locations are areas where flammable vapors are or may be present. Electrical equipment installed in hazardous (classified) locations must be listed for use in these atmospheres so it does not become an ignition source. **[Ref. 1504.6.2]**

External luminaires are attached to the walls or the ceiling of a spraying area. (See Figure 12-11) The luminaires are designed for the replacement of the lamps outside of the spray booth so as to maintain the ordinary hazard electrical classification. **[Ref. 1504.6.2.2]**

Integral luminaires are designed and listed for repair inside of a flammable vapor area. (See Figure 12-12) The fire code requires lamps listed for Class I, Division 2 locations for flammable vapor areas and Class II, Division 2 locations when the luminaire is used for the illumination of powder coating areas. The luminaire also must be evaluated and listed to ensure that residual deposits cannot be heated and ignited by radiant energy from the lamps. **[Ref. 1504.6.2.3]**

FIGURE 12-11 External luminaire. All of the electrical components are located outside of the boundary of the hazardous (classified) area

FIGURE 12-12 Integral luminaire *(Courtesy of Cooper Industries—Crouse Hinds Electrical Division, Syracuse, NY)*

INTERLOCKS

A properly designed mechanical ventilation system includes an interlock with the spraying equipment. The interlock is arranged so the spraying apparatus will not operate if the mechanical ventilation system is not operating. **[Ref. 1504.7.1]**

There are a number of methods that may be used to satisfy this provision. The most common method is to install an electric solenoid valve in the compressed air piping supplying the spraying equipment. (See Figure 12-13) The solenoid valve should be located so it is not within the hazardous (classified) electrical boundary, although valves are available that are listed for use in Class I, Division 2 areas. The solenoid valve is wired into the mechanical ventilation fan motor. When the motor is energized the solenoid valve opens, allowing the flow of compressed air to the spraying equipment.

Another method for satisfying the interlock provision is installing a differential pressure switch which is designed to measure pressure within two areas such as the spray space and the exhaust plenum. (See Figure 12-14) If the pressure exceeds a given value—generally measured in inches of water column—the pressure switch operates and allows the exhaust fan to be energized. The pressure differential switch is considered a more reliable method of interlocking spray equipment with the mechanical ventilation system because the fan will not operate if the paint filters are excessively loaded with overspray. Many spray booth fans are driven by belt drives and a fan may require two or three belts. If a fan belt breaks, the differential pressure will be lower than the operating range of the pressure switch, preventing the operation of the fan motor.

FIGURE 12-13 An electric solenoid valve installed in the compressed air piping

FIRE PROTECTION

Spray booths and spraying rooms are required to be equipped with an automatic fire-extinguishing system. The IFC requires that the fire extinguishing system area of protection include exhaust plenums, exhaust ducts, and both sides of dry filters when they are used. [Ref. 1504.4]

When an automatic sprinkler system is used, NFPA 33 requires the system be designed for an Extra Hazard Group 2 discharge density. NFPA 33 also requires the water supply for the sprinklers installed in the spray booth be controlled by a separate, listed control valve that is accessible from the floor level. NFPA 33 contains specific requirements for the installation of automatic sprinklers in the exhaust ducts and stacks, including maximum spacing requirements and minimum discharge flow rates. [Ref. 903.2.11.6]

FIGURE 12-14 Differential pressure switch

Dry chemical alternative fire extinguishing systems are commonly used for the protection of spray booths or rooms. In sprinklered buildings, dry chemical fire extinguishing systems may be used to satisfy the IFC fire protection requirement. The design of dry chemical extinguishing systems must comply with NFPA 17, *Standard for Dry Chemical Extinguishing Systems.* [Ref. 904.6]

NFPA 17 permits the use of engineered or pre-engineered dry chemical fire extinguishing systems. (See Figure 12-15) Total flooding systems are specified for the protection of spraying areas. A total flooding system is designed to discharge the fire-extinguishing agent into

FIGURE 12-15 Pre-engineered dry chemical alternative fire extinguishing system protecting a spray booth *(Courtesy of TYCO/Ansul Inc., Marinette, WI)*

an enclosure that surrounds the hazard. In a spray booth or room, the interior of the spray space is the hazard being protected. Fire protection system manufacturers offer a variety of pre-engineered total flooding designs for the protection of open-face, cross-draft, and down-draft spray booths.

CHAPTER 13

Motor Fuel-Dispensing Facilities and Repair Garages

Motor fuel-dispensing facilities and repair garages are closely regulated because they allow the general population to handle flammable and combustible liquids and flammable gases that are more hazardous than the public may realize. All but two U.S. states allow the public to perform their own fuel dispensing at public-accessible fueling facilities. The dispenser is often the point of sale and source of revenue. To minimize the risk of fire or unauthorized discharge of hazardous materials, IFC Chapter 22 sets forth requirements based on the hazards of the various fuels available in the market place. Regardless of the fuel, dispensing equipment and components accessible and used by the motoring public must be listed by a nationally recognized testing laboratory.

TABLE 13-1 Applicable IFC requirements and NFPA standards by fuel

Stored and dispensed fuel	Applicable IFC requirements	Applicable NFPA standard(s)
Gasoline, Diesel Fuel, Gasoline/ Ethanol mixtures	Section 2205; Chapter 34	NFPA 30; NFPA 30A
Compressed Natural Gas	Section 2208; Chapter 30; International Fuel Gas Code	NFPA 52
Liquefied Natural Gas	Chapter 35	NFPA 55; NFPA 57; NFPA 59A
Liquefied Petroleum Gas	Section 2207; Chapter 38	NFPA 58
Compressed Hydrogen	Section 2209; Section 3003	NFPA 55
Liquefied Hydrogen	Section 2209; Section 3203	NFPA 55

APPLICABLE REQUIREMENTS BY FUEL

IFC Chapter 22 provisions are based on the type and hazards of the fuel. They are all flammable or combustible and they can be a liquid or gas, and the gases can be compressed or cryogenic (refrigerated gases). The IFC prescribes specific requirements because of the characteristics of the particular fuel and its hazards.

Chapter 22 addresses storage and dispensing of fuels like unleaded gasoline, alcohol-gasoline mixtures, and petroleum distillates, such as number two diesel fuel, kerosene, and biodiesel. It also addresses liquefied petroleum gases (LP-Gas), and natural gas (methane) and hydrogen stored as compressed gases or cryogenic fluids (which are refrigerated gases). The provisions for these fuels are applied in conjunction with the hazardous material requirements in IFC Chapters 30 through 44 and the applicable NFPA standards. Regardless of the fuel and its physical state, all motor fuel-dispensing stations must comply with the general provisions in Section 2201 and the adopted NFPA standards indicated in Table 13-1.

DISPENSING OPERATIONS AND DEVICES—ALL FUELS

Dispensing is the *pouring or transferring of any material from a container, tank, or similar vessel whereby vapors, dusts, fumes, mists, or gases are liberated to the atmosphere.* Dispensing operations under the IFC may be attended or unattended. Attended dispensing occur where an individual at the motor fuel-dispensing station supervises and performs the transfer of fuel. Unattended, or self-service operations are the more common operation because an attendant is not involved and dispensing is performed by the person purchasing the fuel. [Ref. 2702.1]

At attended dispensing operations, at least one responsible person is on site who supervises, controls and observes the fuel dispensing operations. This individual is responsible for ensuring containers filled with fuel comply with the IFC Chapter 34 requirements. The attendant is responsible for

FIGURE 13-1 Attended motor-fuel dispensing

FIGURE 13-2 Unattended motor-fuel dispensing

FIGURE 13-3 Dispenser operating instructions and warning statements are required for attended and unattended motor-fuel dispensing

controlling spills and is trained in the use of portable fire extinguishers. The attendant must be able to communicate with persons in the dispensing area and have access to the emergency controls to stop dispensing if an incident occurs. (See Figure 13-1) [Ref. 2304.2]

Unattended fuel dispensing is self-service dispensing and commonly the dispenser serves as the point of sale. (See Figure 13-2) The IFC requires the owner perform a daily reconciliation of fuel sales as well as an inspection of dispensing equipment. The daily reconciliation verifies the fuel storage tanks are not leaking or otherwise losing product. Each dispenser requires operating instructions and identification of the location of emergency controls that can be used to stop dispensing operations, as well as statements which address dispensing of fuels into unauthorized containers. The signs inform individuals dispensing fuel of the actions they need to take in the event of a fire, spill, or release. (See Figure 13-3) An on site means of communicating with the fire department is required. [Ref. 2204.3 and 2205.6]

Dispensing devices must be properly located in relation to property lines, buildings and their openings, and fixed sources of ignition. Dispensers are be located so the vehicle receiving fuel and the dispensing nozzle are located on the same property. Regardless of the fuel, all dispensers require a clearly identified emergency disconnect switch that stops dispensing upon activation. (See Figure 13-4) Switches can be located indoors or outdoors. Outdoor switches must be a minimum of 20 feet but not more than 100 feet from a dispenser. The 20-foot minimum separation protects individuals who are self-dispensing by removing the person from potential harm if a fuel spill is ignited. [Ref. 2203.1 and 2203.2]

The IFC also establishes operational requirements for tank filling, maintenance of dispensers, and the control of spills and ignition sources. (See Figure 13-5) Before a storage tank is filled, the tank vehicle driver is required to gauge (measure) the tank and determine that the available

FIGURE 13-4 Emergency disconnect switch

FIGURE 13-5 Connections from a tank vehicle to a storage tank must be liquid and vapor tight

FIGURE 13-6 Devices such as these vapor and liquid leak detectors in a below-grade vaulted storage tank must be annually tested

capacity before transferring product. The liquid transfer and vapor recovery connections on storage tanks with a volume of more than 1,000 gallons must be liquid and vapor tight. When liquid is pumped into aboveground storage tanks, the tank vehicle must be located at least 15 feet from a tank receiving Class II and IIIA combustible liquids and 25 feet from a tank receiving a Class I flammable liquid. [Ref. 2205.1]

Dispensing equipment must be properly maintained so it does not become a source of ignition or a leak. Any equipment that is leaking should be removed from service as it is a well known fact that leaks never become smaller. Emergency shutoff valves and liquid leak detectors require an annual functional test. (See Figure 13-6) When repairs are performed on dispensing devices, the IFC requires the electrical power to the dispenser and its source pump be disconnected, the dispenser emergency shutoff valve is closed, and a minimum 12-foot exclusion zone be established. Only persons knowledgeable in the performing the repairs are allowed within the work zone. [Ref. 2205.2.3]

Portable fire extinguishers are required near dispensers, pumps, and storage tank fill connections. Minimum 2-A:20-B:C portable fire extinguishers are required within 75-feet of the indicated components. Portable fire extinguishers must be installed in accordance with IFC Section 906 and NFPA 10. (See Figure 13-7) [Ref. 2205.5]

FLAMMABLE AND COMBUSTIBLE LIQUID MOTOR-FUEL DISPENSING

Unleaded gasoline, biodiesel, number two diesel fuel, and alcohol blended gasoline are commonly dispensed liquid fuels. Section 2206 addresses the storage, transfer, dispensing, and vapor recovery of these liquids. Because stationary storage tanks are used, their installation must comply with the requirements in IFC Chapter 34 and NFPA 30 *Flammable and Combustible Liquids Code*—however, the type of tank selected and its siting requirements must comply with the requirements in IFC Chapter 22,

NFPA 30A *Code for Motor Fuel-Dispensing Facilities and Repair Garages*, and Chapter 34 (A complete review of storage tank design and installation requirements are found in Chapter 18 of this text). **[Ref. 2206.2]**

A common method of storing fuels is underground storage tanks (USTs). USTs are desired by many petroleum marketers because they eliminate the fire hazard of aboveground storage and reduce the needed land area since the tank is buried. Because the tank is underground, leaks cannot be readily observed. IFC Chapter 34 requires a means of leak detection for USTs storing Class I, II, and IIIA flammable and combustible liquids. However, the IFC imposes an additional requirement for daily product reconciliation between product sold, loss due to spills or evaporation, product received, and the available inventory in the UST. A daily reconciliation is required for each storage tank and any consistent or accidental loss of petroleum product must be reported to the fire code official. In many cases, this reconciliation is performed automatically by the tank's inventory control equipment. (See Figure 13-8) **[Ref. 2206.2.1.1]**

Because of the general public's proximity to a fuel dispenser during unattended dispensing, and the potential for vehicular impact, the IFC and NFPA 30A requirements for aboveground storage tanks (AST) supplying motor fuel dispensers are more restrictive than the requirements in IFC Chapter 34. These increased requirements are justified because of the increased potential for fire or explosion.

Class I flammable liquids, such as unleaded gasoline or alcohol-based fuels formulated with 85% ethanol/15% gasoline must be stored in a listed protected aboveground storage tank (PAST). (See Figure 13-9) This particular type of AST is listed to the requirements of UL 2085 *Standard for Safety Protected Aboveground Tanks for Flammable and Combustible Liquids* and is designed to resist damage from a hydrocarbon pool fire. The tank is constructed with integral secondary containment, bullet-resistance, and vehicle impact protection. Class II and IIIA combustible liquids such as number 2 diesel fuel and biodiesel also must be stored in a PAST—however, the fire code official can permit the use of other ASTs for these liquids, since they have a higher flash point temperature, making them more difficult to ignite. The maximum volume for any AST at a publicly-accessible fueling facility is 12,000 gallons, and the aggregate volume at the site is 48,000 gallons. All tanks must be located in accordance with the requirements in Table 2206.2.3, which are more restrictive than the tank siting requirements in Chapter 34. **[Ref. 2206.2.3]**

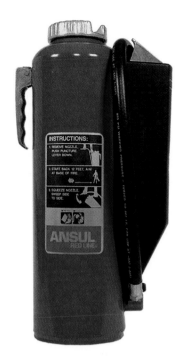

FIGURE 13-7 Portable fire extinguishers serving dispensers, pumps, and tank fill connections require a minimum 20-B class B fire hazard rating *(Courtesy of TYCO/Ansul, Marinette, WI)*

FIGURE 13-8 Electronic UST inventory and leak detection device

Another method that can increase the stored volume is a vaulted tank. Vaulted tanks can be installed above-grade or below-grade. A vaulted tank utilizes a conventional, non-insulated AST. The storage tank is placed inside of a concrete vault which has a concrete cover.

FIGURE 13-9 A protected aboveground storage tank

FIGURE 13-10 Interior of a below grade vaulted aboveground storage tank

FIGURE 13-11 Overfill **prevention device** *(Courtesy of OPW Fueling Components, Dublin, OH)*

Such a design allows for an AST to be installed below-grade, which can benefit the owner because these tanks are exempt from required payments into the jurisdiction's leaking storage tank fund. The liquid tight vault serves as secondary containment if a leak occurs in the primary tank. Vaulted tanks are listed as meeting UL 2245, *Below-Grade Vaults for Flammable Liquid Storage Tanks* and require a mechanical ventilation system, overfill prevention device, and a means of vapor and liquid leak detection. Vaulted ASTs are permitted to store up to 15,000 gallons of Class I, II, and IIIA liquids. (See Figure 13-10) **[Ref. 2206.2.4]**

To reduce the potential for leaks, the IFC has requirements for the design of the piping and valve systems ensure that if a leak occurs, the amount released is minimized. These provisions are used in conjunction with the requirements of IFC Chapter 34 and NFPA 30A.

All tank openings are required to be through the top of the storage tank. By requiring openings at the top of the tank, all liquids must be pumped to the dispenser, which is safer than using a gravity delivery system. Such a system ensures that liquid is below any openings, limiting the potential for leaks. To prevent the overfilling of an AST, the IFC requires the installation of an overfill prevention device. These devices are installed in the tank fill pipe connection and are designed to limit the flow of liquid when the tank is filled to 90% of its volume. When the tank is filled to 95% of its volume, the overfill prevention device closes and prevents the delivery of any additional petroleum product. (See Figure 13-11) **[Ref. 2206.6.2.3]**

Because the IFC requires all openings for tanks to be through the top of the tank, transfer pumps are required. Pumps can be installed at the top of the tank or can be installed at the dispenser. When a pump is installed at the dispenser, the potential for product siphoning can occur. If the pump is operated and the piping has a leak, the potential exists for the negative pressure to siphon liquid from the storage tank, which can increase the size of the release. When such designs are used, the IFC requires the installation of an anti-siphon valve. This can be a solenoid valve that opens when the dispenser is activated or the installation of a valve that is designed to prevent siphoning of product in the dispenser supply pipe. (See Figure 13-12) **[Ref. 2206.2.4]**

A fuel dispenser is the point of use by the consumer. Because the predominance of fuel dispensing is performed in an unattended setting, the IFC specifies requirements to minimize the potential of a spill and fire. These requirements address the method of fuel delivery, mounting and construction of dispensers, and the hose and nozzles.

All components of a flammable or combustible liquid dispensing system are required to be listed by a nationally recognized testing laboratory. The IFC requires listed electrical equipment, dispensers, hoses, nozzles, and submersible or subsurface pumps used for the movement and dispensing of liquid. Listing of equipment ensures that it has been evaluated for the fuel being dispensed and is constructed and assembled using materials compatible with the fuel. (See Figure 13-13) **[Ref. 2206.7.1]**

A major concern with any dispenser is vehicular impact. Dispensers are equipped with valves and fittings that prevent the release of flammable and combustible liquids if it is impacted by a vehicle. The IFC requires the installation of a dispenser emergency shutoff valve at the base of the dispenser in each product line that supplies a dispenser. The valve is equipped with a low melt point fusible link so in the event of a fire at the dispenser, the link will fail and close the valve.

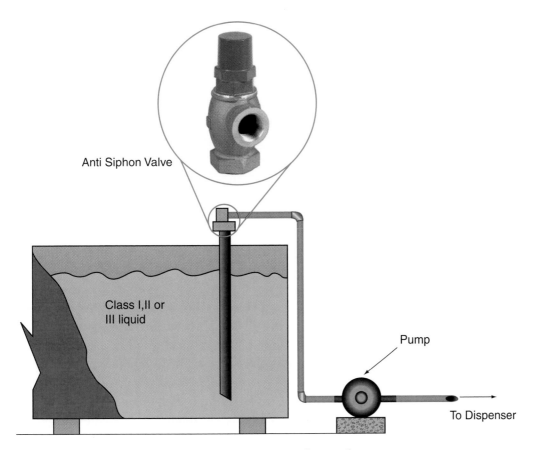

Anti Siphon Valve

Class I,II or III liquid

Pump

To Dispenser

FIGURE 13-12 Anti-siphon valve

FIGURE 13-13 Variable frequency drive pump controllers. These devices control the flow rate of pumps supply dispensers. Because they are part of the dispensing system, these controllers must be listed to comply with the IFC

FIGURE 13-14 Dispenser emergency shutoff valve
(Courtesy of OPW Fueling Components, Dublin, OH)

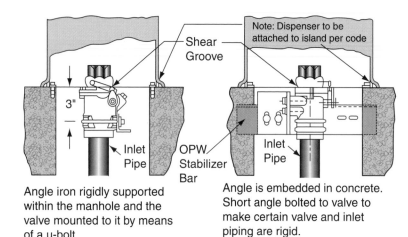

Angle iron rigidly supported within the manhole and the valve mounted to it by means of a u-bolt.

Angle is embedded in concrete. Short angle bolted to valve to make certain valve and inlet piping are rigid.

FIGURE 13-15 Proper installation of the dispenser emergency valve should include verification of the location of shear section in relation to the foundation
(Courtesy of OPW Fueling Components, Dublin, OH)

Dispenser emergency shutoff valves are required to be functionally tested annually. (See Figure 13-14) **[Ref. 2205.2.4 and 2206.7.4]**

A dispenser emergency shutoff valve has a shear section that will separate from the main body of the valve attached to the dispenser base. Separation of the shear section mechanically stops the flow of liquid. Positioning of the shear section is important, because it must be parallel to the dispenser foundation and properly located so the shear section can effectively operate when impacted. The IFC requires it be installed within ½ inch of the top of the island—equally important is that the valve be installed in accordance with its listing. (See Figure 13-15) **[Ref. 2206.7.4]**

The U.S. Environmental Protection Agency can mandate the installation of vapor-recovery and vapor-processing systems to capture vapor that can be released during the dispensing of unleaded gasoline and ethanol/gasoline mixtures. Vapor recovery piping is installed as a part of the dispenser so a dispenser emergency shutoff valve is required. (See Figure 13-16) **[Ref. 2206.7.4]**

Another safety component installed on a dispenser is an emergency breakaway device that is designed to safely separate the dispensing hose if a vehicle drives away when the fuel delivery nozzle is attached to the vehicle. (See Figure 13-17) An emergency breakaway device is required

FIGURE 13-16 A dispenser emergency valve for vapor recovery service *(Courtesy of OPW Fueling Components, Dublin, OH)*

FIGURE 13-17 An emergency breakaway device

on all dispensers conveying Class I and II liquids. When operated, the emergency breakaway device will contain liquid on both sides of the hose that is separated. **[Ref. 2206.7.5.1]**

The point of use for motor-fuel dispensers is the fuel delivery nozzle. Fuel delivery nozzles dispensing Class I, II, or IIIA liquids are automatic closing. The fire code official can dictate when a latch open device is or is not allowed. Some code officials do not allow latch open devices because it requires the person dispensing the fuel to be in constant attendance of the delivery nozzle. Conversely, other fire code officials require latch open devices because it prevents the person dispensing fuel from inserting objects between the nozzle body and valve, such as vehicle fuel caps or wallets. The IFC requires an integral latch open device. (See Figure 13-18) **[Ref. 2206.7.6.1]**

Fuel delivery nozzles are designed so they will only allow dispensing when the dispenser hose is pressurized with liquid. Loss of pressure, such as turning off a pump, will cause the nozzle valve to close. The nozzle must be manually operated to ensure it is closed before dispensing can be resumed. Its design must have a feature or component that retains the nozzle in the vehicle fill pipe while dispensing is underway. **[Ref. 2206.7.6]**

Alcohol-blended fuel is defined in Section 2202.1; the definition specifies common fuel formulations. Ethanol is a water-miscible solvent classified as a Class I-B flammable liquid because of its closed cup flash point temperature of 55°F and a boiling point temperature of 173°F. Ethanol (designated as "E") is blended with

FIGURE 13-18 A listed fuel delivery nozzle with a latch open device

TABLE 13-2 Flammability ranges of unleaded gasoline and alcohol-blended fuel

Fuel	Lower flammable limit (percent volume in air)	Upper flammable limit (percent volume in air)
Unleaded gasoline	1.4%	7.6%
E85 alcohol-blended fuel	1.4%	19.0%

gasoline to make up two different formulations of alcohol-blended fuels: E10, which is a blend of 10% ethanol with 90% unleaded gasoline, and E85, which is a blend of 85% ethanol with 15% gasoline. E85 alcohol-blended fuel has a closed cup flash point temperature ranging from −20 to −4°F, which is higher than unleaded gasoline, which has a −45°F flash point temperature. The flammable range of E85 alcohol-blended fuel is much greater than that of unleaded gasoline as indicated in Table 13-2.

The IFC requires listed components for the dispensing of alcohol-blended fuels, such as nozzles, hoses, breakaway connections, or any other component that will be wetted by the fuel. Compared with gasoline, ethanol and ethanol-blended fuels have increased electrical conductivity, which can affect material compatibility as a result of increased corrosion. In this context, *compatibility* is defined as the ability of two or more substances to maintain their respective physical and chemical properties upon contact with one another for the design life of the storage and dispensing system under conditions likely to be encountered. Alcohol-blend fuels will aggressively attack and cause premature failure of components constructed of metals, alloys, and plastics that are not chemically compatible with ethyl alcohol. Components constructed of soft metals, including zinc, brass, aluminum, or lead, are chemically incompatible with E85 alcohol-blended fuel. Unplated carbon steel, stainless steel, and bronze are resistant to ethanol. Nonmetallic materials that are compatible with alcohol-blended fuels include neoprene rubber, polypropylene, nitrile plastic, and polytetrafluoroethylene (registered under the trademark Teflon). **[Ref. 2206.8.1]**

If a motor vehicle fuel-dispensing facility switches existing storage tanks and dispensing equipment to an alcohol-formulated fuel, the conversion is subject to a review and approval by the fire code official. Ethanol is an extremely water-miscible liquid, and if equipment is not properly cleaned or prepared for the storage of alcohol-blended fuels, the ethanol will absorb any water and contaminate the fuel. If the facility stores petroleum fuels in underground storage tanks constructed of fiberglass-reinforced plastic, the tank may or not be compatible with E85 fuel. Since the mid-1980s, all fiberglass-reinforced plastic underground storage tanks constructed with integral secondary containment (such as a double-wall) are compatible with 100% ethyl alcohol. In each case it is important that the original tank installation records be reviewed and any questions concerning the compatibility of the tank with the fuel be referred to the manufacturer. Automatic tank liquid level gauges using capacitance probes will not work with alcohol-blended fuels. Therefore, part of the approval process should include a review of manufacturer data sheets or listings to confirm the component's compatibility with ethanol. (See Figure 13-19) **[Ref. 2206.8.2]**

FIGURE 13-19 An E-85 alcohol-unleaded gasoline dispenser

LIQUEFIED PETROLEUM GAS DISPENSING

Liquefied petroleum gas (LP-Gas) is a flammable liquefied compressed gas used as a motor vehicle fuel. In comparison to conventional fuels like unleaded gasoline and diesel fuel, LP-Gas is the third most popular fuel used in the United States. Being a liquefied compressed gas, LP-Gas is stored and dispensed as a liquid from a stationary container or tank into the vehicle fuel tank by way of a pump and piping. Before it can be used as a motor fuel, the vehicle must vaporize the liquid and convert it to a gas for carburetion.

LP-Gas storage and dispensing must comply with the requirement in IFC Chapters 22, 30, and 38 and the requirements in NFPA 58, *Liquefied Petroleum Gas Code.* The stationary LP-Gas container and its piping are sited and installed in accordance with NFPA 58 and Chapter 38. The dispenser, hose, and nozzle are installed in accordance with all of the requirements in Chapter 22 and the NFPA 58 requirements for dispensers. **[Ref. 2207.5]**

(See Figure 13-20) Listed hoses, hose connectors including break-away connections, vehicle fuel connectors (nozzles), dispensers, and LP-Gas pumps are required. (See Figure 13-20) The storage container, piping, pressure relief devices, and pressure regulators must be approved by the fire code official. **[Ref. 2207.2.2]**

The IFC requires the point of transfer be separated from buildings, property line, streets, and public ways. Point of transfer is the location where dispensing occurs: this is where the connection between the vehicle and the dispensing nozzle are made and broken. (See Figure 13-21) During any LP-Gas dispensing operation, a very small volume of liquid is trapped between the nozzle and vehicle fill connection. Disconnection

FIGURE 13-20 This LP-Gas pump is required by the IFC to be listed

FIGURE 13-21 The point of transfer is measured from the location where connections are made or disconnected

of the nozzle will cause this liquid to vaporize and release into the atmosphere. Therefore, the separation distances should be measured from the dispenser location. A minimum separation of 25 feet is required between the point of transfer and buildings with combustible exterior walls or noncombustible walls that have less than 1-hour fire resistance, lot lines on property that can be built on, streets, sidewalks, and railroads. The point of transfer separation is reduced to 10 feet when buildings have noncombustible exterior walls with a fire-resistance rating of 1 hour or more. [Ref. 2207.4]

NFPA 58 and the IFC set forth the requirements for the installation of LP-Gas dispensers. An excess flow control valve and a manual shutoff valve are required in the piping between the transfer pump and the dispenser inlet. An excess flow control is designed to stop the flow of liquid LP-Gas automatically in the event a pressurized pipe fails. The valve is sized based on the diameter of the pipe and the liquid flow rate. Therefore, if the diameter of the piping changes as part of the design or installation, an excess flow control valve is required at the point where the pipe diameter changes. Excess flow control valves are not required by the IFC or NFPA 58 to be listed. (See Figure 13-22) [Ref. 2207.5.1]

A second excess flow control valve is required at the connection of the dispenser hose to the liquid LP-Gas piping. This valve prevents the release of liquid LP-Gas if the dispensing hose ruptures. An excess flow control valve is not required when a differential backpressure valve is used. (See Figure 13-23) A differential backpressure valve operates on the concept of constant flow and pressure. In the event of a pipe or hose failure, the flow rate increases, which in turn causes the valve to close because of the pressure increase between the dispenser and tank. [Ref. 2207.5.1]

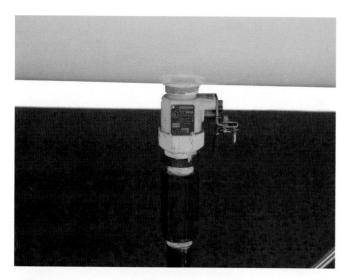

FIGURE 13-22 An internal excess flow control valve. Because of the size of the LP-gas container, it is equipped with a fusible link that closes the valve in the event of a fire

FIGURE 13-23 A differential backpressure valve

Dispenser hoses are listed for LP-Gas service and equipped with a listed quick-action shutoff valve. This valve is designed such that liquid LP-Gas cannot flow until a positive mechanical connection is made at the vehicle fill connection. (See Figure 13-24) While not specified by the IFC, NFPA 58 requires the installation of a listed emergency breakaway connection on the dispenser hose. [Ref. 2207.5.1 and 2207.5.2]

All liquefied compressed gases exhibit fairly high expansion ratios at normal temperatures. Because LP-Gas is a mixture of ethane, propane and butane, the expansion can vary, but it is generally about 36.4 cubic feet/gallon. At 70°F LP-Gas has a vapor pressure of 145 PSIG. Therefore, if the liquid is trapped in a closed pipe or hose, the pressure can increase as the pipe or hose is warmed to a point where catastrophic fail-

FIGURE 13-24 A listed quick-action dispenser nozzle for LP-Gas

ure can occur. To prevent such a failure, the IFC requires the installation of one or more hydrostatic pressure relief valves. This valve is designed to safely vent the LP-Gas once the pressure reaches a preset point. When the pressure is relieved, the valve closes. A hydrostatic pressure relief valve is required by NFPA 58 in any section of pipe or hose where liquid LP-gas can be trapped (See Figure 13-25). This includes the LP-gas dispenser hose because liquid can be trapped between the dispenser valve and the LP-Gas pump. [Ref. 2207.5.2]

FIGURE 13-25 Hydrostatic pressure relief valves

High-Piled Combustible Storage

With the cost of land increasing and changes in technologies that allow for rapid fulfillment of a customer's request for manufacturing and consumer goods, high-piled combustible storage is a preferred method of storage in storage (Group S) and many mercantile (Group M) occupancies. Many retailers for home improvement, consumer electronics, and consumer soft goods such as furniture, clothing, food, and beverages construct large Group M occupancies that use high-piled combustible storage to satisfy their customer demands and to control costs.

High-piled combustible storage allows for a greater amount of combustible materials within a given floor area. This increased fuel load increases the potential fire dollar loss for each square foot of floor area; property losses resulting from smoke and fire damage of goods within the building in many cases can surpass the construction cost of the building.

It is challenging in buildings housing high-piled combustible storage to verify the adequacy of the automatic sprinkler system. In the past 15 years, the fire protection engineering community has witnessed the introduction of a variety of new automatic sprinklers that are designed specifically for high-piled combustible storage. The available methods of design have also increased and the NFPA 13 requirements for automatic sprinkler protection are continuing to be extensively modified.

WHAT IS HIGH-PILED COMBUSTIBLE STORAGE?

High-Piled Combustible Storage is the *storage of combustible materials in closely packed piles or combustible materials on pallets, in racks or on shelves where the top of storage is greater than 12 feet in height. When required by the fire code official, high-piled combustible storage also includes certain high-hazard commodities, such as rubber tires, Group A plastics, flammable liquids, idle pallets and similar commodities, where the top of storage is greater than 6 feet in height.* [Ref. 2302.1]

The provisions in IFC Chapter 23 become applicable when the goods are stored in piles, on pallets, in racks, or on shelves and the height of storage is greater than 12 feet. The 12-foot limit is based on fire testing that was performed in the 1960s and 1970s. The tests revealed that when goods were stored more than 12 feet in height, the rate of fire spread and heat release dramatically increased to the point that a sprinkler system designed for an Ordinary Hazard in accordance with NFPA 13 could not control the fire.

The threshold for application of Chapter 23 for commodities classified as High-Hazard is 6 feet. They include rubber tires, highly combustible and fast burning (termed Group A) plastics, flammable and combustible liquids, idle pallets, and alcohol or hydrocarbon formulated aerosols. NFPA 13, Chapter 12 contains automatic sprinkler design criteria for storage of High Hazard commodities including rubber tires, Group A plastics, idle wood and plastic pallets, and rolled paper. High hazard commodities almost always require a specialized automatic sprinkler system that is designed to control or suppress fires involving these goods. (See Figure 14-1) [Ref. 903.3.1.1]

FIGURE 14-1 Hydrocarbon formulated aerosols represent a very significant challenge in the design of the automatic sprinkler system and are treated as a High Hazard Commodity by the IFC

COMMODITY CLASSIFICATION

To determine if the requirements of IFC Chapter 23 are applicable, the stored commodities must be classified in accordance with Section 2303, the height of storage must be established, and the high-piled combustible storage area be designated. The commodity classification and the height of storage will influence the design of the automatic sprinkler system when the system is required.

Commodities are classified based on an estimation and comparison of the heat release rate of typical products in the category. The overall fire hazard of a commodity is a function of its heat release rate. Heat release rate is measured in BTU/minute (KW). The heat release rate is the product of the heat of combustion measured in BTU/pound (kJ/kG), and the burning rate is measured in pounds/minute (kg/second). The higher a commodity's heat of combustion and heat release rate, the higher the commodity classification. The higher a commodity's classification, the more difficult this material is to control and extinguish when it is involved in fire.

The IFC divides commodities into five different classes:

- Class I,
- Class II,
- Class III,
- Class IV, and
- High-Hazard.

Figure 14-2 illustrates that a Class I commodity is considered to have the lowest heat release rate, while a High-Hazard commodity has the highest heat release rate.

Before making a determination of a commodity's hazard classification, it is important to review the definition of commodity: *A combination of products, packing materials and containers.* Flammability properties of the product, packaging, and containers materials need to be evaluated in establishing appropriate fire protection. The commodity classification is not limited to the packaged product but also includes the packing material and the material of construction for the container. Packing materials can include a large amount or volume of plastics, which generally exhibit higher heat of combustion when compared to other materials constructed of noncombustible and combustible materials. Consider the definition of a Class II commodity and compare it to a Class I commodity. A Class II commodity includes Class I products that are packaged in slatted wooden crates, solid wooden boxes, multiple layer paper or fiberboard cartons, or equivalent combustible packaging materials with or without pallets. [Ref. 2302.1, 2303.2, and 2303.3]

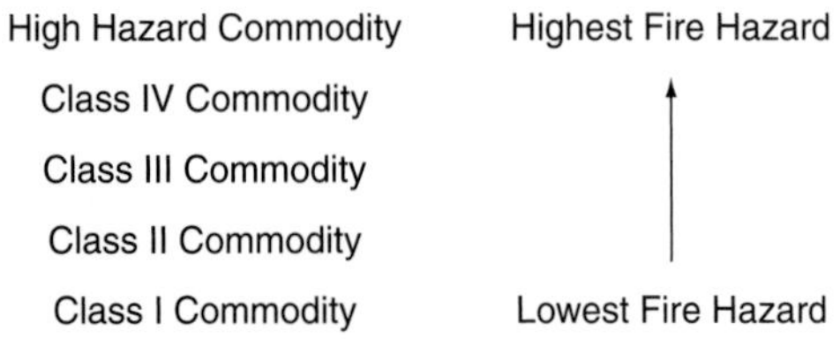

The IFC requires that the physical form of the commodity be evaluated when assigning the commodity classification. The physical form of a commodity is commonly referred to as its geometry which must be considered when classifying

High Hazard Commodity — Highest Fire Hazard

Class IV Commodity

Class III Commodity

Class II Commodity

Class I Commodity — Lowest Fire Hazard

FIGURE 14-2 **The higher the commodity class, the greater the challenge in the design of the automatic sprinkler system**

commodities. Materials will burn differently based on their geometry. For example, consider dimensional wood lumber. Because of its compacted mass, it is difficult to burn when stored horizontally. However, if the same dimensional lumber is assembled into uniform shapes to create a conventional pallet, more exposed surface area is created. The result is that a material that is normally classified as a Class III commodity by IFC Section 2303.4 is now classified as a High-Hazard commodity by Section 2303.6 simply by changing the geometry of the wood. (See Figure 14-3) **[Ref. 2303.1]**

A review of the Class III, IV, and High-Hazard commodity classifications finds they include plastics. Plastics are classified separately in the IFC and NFPA 13 because they have such a wide range of heat of combustion per unit of mass that must be considered when assigning classifications. Because of the large number of plastics, the complexity of their nomenclature and the ease of changing burning characteristics with additives, careful consideration of the resins is necessary when classifying plastics. Certain plastics pose a significantly greater hazard than ordinary combustibles; therefore, plastics are classified separately.

The heat release rate (Btu/min or kW) for plastics can be two to three times greater than for a similar arrangement of ordinary combustibles. For example, the heat of combustion of ordinary combustibles, such as wood or paper, generally ranges between 6,000 and 8,000 Btu/lb. The heat of combustion for plastics generally ranges between 12,000 and 20,000 Btu/lb. The burning rate of a commodity is dependent on many things, but plastic materials generally exhibit higher maximum burning rates than similarly arranged ordinary combustibles.

FIGURE 14-3 The geometry influences the commodity classification. The dimensional wood on the left is a Class III commodity, while the finished wood fence sections on the right is a High-Hazard Commodity

Plastics are classified as either Group A, Group B, or Group C. Group A plastics represent the most challenging from a fire protection viewpoint while Group C plastics are the least challenging. **[Ref. 2303.7]**

The geometry of the plastic is important. The geometry will influence the ease of ignition and the burning rate of plastics. Plastics have three basic geometric forms:

- Expanded
- Unexpanded
- Free-flowing

Expanded plastics are generally a low-density product and are commonly called "foam plastics" such as polystyrene foam coffee cups or packaging material, and polyethylene and polypropylene foam sheeting. Expanded plastics have a cellular structure made up of many small air pockets and voids. The pockets increase the available surface area and promote easy ignition and increased vertical flame spread. (See Figure 14-4)

Unexpanded plastics have a higher density when compared to expanded plastics and may or may not have rigidity. Plastic films are classified as unexpanded plastic as well as plastic sheets. They are a solid material and can include goods such as toys, tote bins, and containers. In comparison to expanded plastics, unexpanded plastics are less hazardous because of the reduced surface area the heat of combustion is generally the same because this value is influenced by the resin formulation. (See Figure 14-5)

Free-flowing plastics are very small plastic items such as bottle caps, hypodermic needle plungers, granular or flake plastics, or powdered plastics. Free-flowing plastics burn less severely when compared to expanded or unexpanded plastics. When a package of free-flowing plastics fails under fire exposure, the goods spill from the container. In rack storage, the items begin to fill the rack flue spaces and slow the spread of a fire. Because of this phenomenon, free-flowing plastics are classified as a Class IV commodity. (See Figure 14-6) **[Ref. 2303.5]**

FIGURE 14-4 Expanded Group A Plastic

FIGURE 14-5 Unexpanded Group A plastic box

HIGH-PILED COMBUSTIBLE STORAGE AREAS

After the individual commodities are classified, the area of high-piled storage is designated. A high-piled storage area is a space within a building which is designated, intended, proposed, or actually used for high-piled combustible storage. The IFC requires that the most challenging commodity be used as the basis for classification of the high-piled storage area—if a storage area has a mix of Class I, III, and IV commodities, the fire protection must be provided for the Class IV hazard level. This designation will not impact the design of fire department access and firefighter access doorways. This designation will impact the design of smoke and heat vents and the automatic sprinkler system. [Ref. 2304.1]

The Class IV commodity designation is the most commonly applied classification and protection application that is assigned to High-Piled Storage Areas. Excluding High-Hazard commodities, this assignment is common because it allows maximum flexibility in the proposed, current, or future use of buildings. Many jurisdictions require the classification and protection of speculation warehouses based on Class IV commodities.

FIGURE 14-6 Free-flowing Group A plastic

After the commodities are classified and storage height and high-piled storage area have been determined and identified, the amount and type of protection required for the facility must be determined. These requirements are set forth in Table 2306.2, which contains the general fire protection and life safety requirements for high-piled combustible storage. This table is as an index to the requirements in IFC Sections 2306.2 through 2306.10.

A basic review of this table reveals that all of the requirements are dependent on the commodity classification and the size of the high-piled storage area. Note that the table is divided into two basic commodity categories: Classes I through IV and High-Hazard. For Class I through IV commodities, there are no real differences in the requirements in the 2009 IFC—once a classification has been assigned to goods, materials, and packaging assigned to one of four of these commodity classes, the requirements of Table 2306.2 are primarily based on the high-piled storage area.

Automatic sprinkler protection is required for high-piled combustible storage areas over 12,000 ft.^{2} housing Class I-IV commodities or storing high hazard commodities in an area greater than 2,500 ft.^{2} For instances where a particular storage area is designed for a lower hazard commodity and a higher hazard commodity is introduced, Section 2306.2 can be used to require improvements to the fire protection systems so the higher hazard commodity can be safely stored. (See Table 14-1) [Ref. Table 2306.2]

For buildings containing high-piled combustible storage that are accessible to the public, such as Mercantile (Group M) occupancies, the area thresholds for requiring automatic sprinkler protection are far less than the base values prescribed in IFC and IBC Chapter 9. The requirement for the installation of an automatic sprinkler system is applicable in public accessible buildings storing Class I-IV commodities

TABLE 14-1 General fire protection and life safety requirements (IFC Table 2306.2)

Commodity class	Size of high-piled storage area[a] (square feet) (see Sections 2306.2 and 2306.4)	All storage areas (See Sections 2306, 2307 and 2308)[b]					Solid-piled storage, shelf storage and palletized storage (see Section 2307.3)		
		Automatic fire-extinguishing system (see Section 2306.4)	Fire detection system (see Section 2306.5)	Building access (see Section 2306.6)	Smoke and heat removal (see Section 2306.7)	Draft curtains (see Section 2306.7)	Maximum pile dimension[c] (feet)	Maximum permissible storage height[d] (feet)	Maximum pile volume (cubic feet)
I–IV	0–500	Not Required[a]	Not Required	Not Required[e]	Not Required	Not Required	Not Required	Not Required	Not Required
	501–2,500	Not Required[a]	Yes[i]	Not Required[e]	Not Required	Not Required	100	40	100,000
	2,501–12,000 Public accessible	Yes	Not Required	Not Required[e]	Not Required	Not Required	100	40	400,000
	2,501–12,000 Nonpublic accessible (Option 1)	Yes	Not Required	Not Required[e]	Not Required	Not Required	100	40	400,000
	2,501–12,000 Nonpublic accessible (Option 2)	Not Required[a]	Yes	Yes	Yes[j]	Yes[j]	100	30[f]	200,000
	12,001–20,000	Yes	Not Required	Yes	Yes[j]	Not Required	100	40	400,000
	20,001–500,000	Yes	Not Required	Yes	Yes[j]	Not Required	100	40	400,000
	Greater than 500,000[s]	Yes	Not Required	Yes	Yes[j]	Not Required	100	40	400,000

High hazard	0–500	Not Required[a]	Not Required	Not Required[e]	Not Required	Not Required	50	Not Required	Not Required
	501–2,500 Public accessible	Yes	Not Required	Not Required[e]	Not Required	Not Required	50	30	75,000
	501–2,500 Nonpublic accessible (Option 1)	Yes	Not Required	Not Required[e]	Not Required	Not Required	50	30	75,000
	501–2,500 Nonpublic accessible (Option 2)	Not Required[a]	Yes	Yes	Yes[j]	Yes[j]	50	20	50,000
	2,501–300,000	Yes	Yes	Yes	Yes[j]	Not Required	50	30	75,000
	300,001–500,000[g, h]	Yes	Not Required	Yes	Yes[j]	Not Required	50	30	75,000

For SI: 1 feet = 304.8 mm, 1 cubic feet = 0.02832 m³, 1 square feet = 0.0929 m².

a. When automatic sprinklers are required for reasons other than those in Chapter 23, the portion of the sprinkler system protecting the high-piled storage area shall be designed and installed in accordance with Sections 2307 and 2308.

b. For aisles, see Section 2306.9.

c. Piles shall be separated by aisles complying with Section 2306.9.

d. For storage in excess of the height indicated, special fire protection shall be provided in accordance with Note g when required by the fire code official. See also Chapters 28 and 34 for special limitations for aerosols and flammable and combustible liquids, respectively.

e. Section 503 shall apply for fire apparatus access.

f. For storage exceeding 30 feet in height, Option 1 shall be used.

g. Special fire protection provisions, including but not limited to: fire protection of exposed steel columns; increased sprinkler density; additional in-rack sprinklers, without associated reductions in ceiling sprinkler density; or additional fire department hose connections, shall be provided when required by the fire code official.

h. High-piled storage areas shall not exceed 500,000 square feet. A 2-hour fire wall constructed in accordance with the *International Building Code* shall be used to divide high-piled storage exceeding 500,000 square feet in area.

i. Not required when an automatic fire-extinguishing system is designed and installed to protect the high-piled storage area in accordance with Sections 2307 and 2308.

j. Not required when storage areas are protected by early suppression fast response (ESFR) sprinkler systems installed in accordance with NFPA 13.

FIGURE 14-7 If high-hazard commodities are introduced into a Group M occupancy, automatic sprinkler protection is required when the storage height is 6 feet or more and the area of storage is greater than 500 square feet

FIGURE 14-8 When a heat detection system is used for the protection of a high-piled storage area, it must be in accordance with NFPA 72

with a high-piled storage area over 2,500 ft.[2] and public accessible high-piled storage areas over 500 ft.[2] housing High Hazard commodities. (See Figure 14-7) By comparison IFC and IBC Section 903.2.6 requires automatic sprinkler protection when the fire area of a Group M occupancy exceeds 12,000 ft.[2]. The intent of this provision is the protection of the building occupants. Group M occupancies can have large occupant loads and fuel packages. Fires involving high-piled combustible storage exhibit high heat release rates. Accordingly, when the Group M occupancy contains high-piled combustible storage, the IFC lowers the area threshold to ensure that design of the automatic sprinkler system can control or suppress an unwanted fire event. **[Ref. Table 2306.2]**

For smaller buildings that are not sprinklered that house high-piled combustible storage, a fire detection system is required when the high-piled combustible storage area contains Class I-IV commodities, has an area between 2,501–12,000 ft.[2], and is not accessible to the public. Buildings not accessible to the public and containing high hazard commodities require a fire detection system when the high-piled storage area is 501–2,500 ft.[2] **[Ref. Table 2306.2]**

When Group S occupancies lack climate control, design professionals usually will select heat detection as a means of complying with the IFC. NFPA 72 contains requirements for the installation of spot-type heat detectors in buildings with high ceilings. The requirements in NFPA 72 reduce the spacing of spot-type heat detectors as the height of the building increases. (See Figure 14-8) NFPA 72 also limits the use of these particular heat detectors to smooth, beam, or sloped ceilings in a building 30 feet or less in height. Above 30 feet in height, NFPA 72 requires a performance design when spot-type heat detectors are specified based on the requirements of NFPA 72 unless a linear cable or pneumatic rate-of-rise heat detection system is used. Above 30 feet, other options could include the use of optical flame detection or linear beam smoke detection systems. **[Ref. Table 2306.2]**

For high-piled combustible storage, the IFC contains specific requirements for fire department access to buildings for manual firefighting and overhaul operations. Sprinklered buildings require building access when the high-piled combustible storage area contains Class I-IV commodities over an area greater than 12,000 ft.[2] or a high-hazard commodity over an area greater than 2,500 ft.[2] In buildings that are not sprinklered, the threshold for building access is 2,501 ft.[2] for Class I-IV commodities and 501 ft.[2] for high-hazard commodities, provided that the storage area is not accessible to the public. Access is required within 150 feet of all portions of the exterior walls of a building that can contain high-piled combustible storage. **[Ref. 2306.6]**

Firefighter access doorways are required depending on the high-piled storage area and the commodities. (See Figure 14-9) Doors must be accessible without use of a ladder and spaced not more than 150 feet apart. The IFC also establishes requirements for minimum door dimensions, locking devices, and limits on the type of door that can be used for access and prohibits the use of roll-up doors unless they are approved by the fire code official. [Ref. 2306.6.1]

Reasons for the prohibition include roll-up doors can be damaged by material handling equipment, which could limit there use by firefighters. Another concern is actually the task of forcing entry through a roll-up door. Generally, the task involves using a motorized saw, cutting an "X" into the door, and pushing the cut metal into the building. Such

FIGURE 14-9 Firefighter access doorway for a building that contains high-piled combustible storage

tasks can lead to firefighter injuries. In 1999, data from the U.S. Fire Administration reported that the rate of firefighter injuries doubled when performing firefighting operations at commercial buildings, and 44% of all injuries were sprains/strains or bleeding/bruises/cuts or wounds[1]. It is for these reasons the fire code prohibits the use of roll-up doors for required access doors unless they are approved by the fire code official. [Ref. 2306.6.1.2]

STORAGE METHODS

A number of storage methods are used in high-piled combustible storage, and these methods are regulated in IFC Sections 2307 through 2310. Each method of storage can present its own fire protection issues. The selected storage method is generally dictated by the stored commodity and economics. Certain storage methods may be limited in height based on the limitations of fire test data.

Solid-pile storage involves commodities that are moved without the use of material handling aid, such as slave or conventional pallets. (See Figure 14-10) Solid-pile storage is commonly found for commodities such as rolled paper, rolled carpet, or materials that are baled. Some lightweight commodities packaged in fiberboard cartons also may be stored using the solid-pile storage method. Materials stored in a solid pile

FIGURE 14-10 Solid pile storage

[1]U.S. Fire Administration – Topical Fire Research Series, *Firefighter Injuries*, Volume 2, Issue 1, July 2001.

are generally capable of supporting the dead load of additional materials added to the pile. Stored commodities are in direct contact with each other. Solid-pile storage is normally limited to storage on a floor of a building. Solid-pile storage has limited or no gaps or spaces that can serve as paths for air movement or fire spread. Fires involving solid pile storage are the easiest to control in comparison to palletized or rack storage because they provide large, exposed surface areas to allow for the application of water.

Pallets are designed to serve as a uniform platform for the packaging of commodities. They are constructed as a flat structure using wood, plastic, metal, or paper. Palletized storage fires are more challenging to control in comparison to solid-pile storage. (See Figure 14-11) The pallet allows for the horizontal spread of fire. By design, pallets have openings to accommodate the mechanical handling equipment. These openings create flue spaces that allow the horizontal spread of fire. Another consideration for fires involving palletized storage is the horizontal flue space created by the inherent design of pallets is shielded from water discharged by an automatic sprinkler system. As a result, pre-wetting cannot occur because of the shielding. Because of these concerns with palletized storage, the IFC limits pile lengths for Class I-IV commodities to a maximum length of 100 feet and high-hazard commodities to a maximum pile length of 50 feet. [Ref. 2307.3]

A variable that can influence the design of the automatic sprinkler system is plastic pallets used as a material handling aid. Pallets, when constructed of combustible materials such as wood or plastic, represent a significant fire threat to a building and a very challenging automatic sprinkler system design. Pallet fires have very high heat release rates because of the large surface area to mass ratio. The IFC has specific requirements for plastic pallets stored inside of a high-piled combustible storage area. The IFC allows the use of and recognizes UL 2335, *Fire Tests of Storage Pallets*, as a basis for testing and listing of reduced-hazard pallets. (See Figure 14-12) Pallets listed to this standard can be

FIGURE 14-11 Palletized storage

FIGURE 14-12 UL 2335 listed plastic pallet
(Courtesy of Rehrig Pacific Co., Los Angeles, CA)

treated as a Class II commodity versus unlisted plastic pallets, which are a high-hazard commodity and require a specially engineered fire protection system. Listed pallets can be identified by the listing mark which is molded into the pallet. [Ref. 2308.2.1]

Rack storage is the most predominant of all the methods of storage regulated by the IFC. Rack storage is commonly used because it can be erected rapidly, it is designed to facilitate a wide range of storage practices, and its height is generally unlimited because of its inherent structural design as a metal frame. Rack storage are designed for mechanical or manual stocking and retrieval. A cursory review of NFPA 13 and Factory Mutual Global Loss Prevention Data Sheets finds that the greatest number of automatic sprinkler design options exists for rack storage.

While it is not defined in the IFC or IBC, a storage rack is a combination of vertical, horizontal, or diagonal structural members designed to support stored materials. (See Figure 14-13) Racks are constructed with or without solid shelves. Racks can be stationary, moveable, or portable. Portable or moveable storage racks represent their own fire protection engineering challenges and are outside the scope of this chapter. There are no provisions in the IFC or NFPA 13 that limit the height of storage racks. It is a common construction practice to use the building storage racks as the load-bearing structural elements for the walls and roof of a building.

Of all of the storage configurations regulated in the fire code, Rack storage of commodities represents the most challenging from a fire protection perspective. Fires involving rack storage (See Figure 14-14) are one of the most challenging fires for an automatic sprinkler system to control or suppress because:

- The rack supports the commodity. This allows the commodity to be preheated and burn on the bottom and all vertical surfaces.
- A storage rack is constructed with air spaces on all sides that allows for rapid vertical and horizontal fire spread.

FIGURE 14-13 Rack storage of commodities

FIGURE 14-14 Fire involving rack storage of motor oil (*Courtesy of International Code Consultants Inc., Austin, TX*)

- The flue spaces between stored commodities allow for the velocity of the fire gases to be increased by compressing them between structural members and the exposed surfaces of the commodities. This increase of the gas velocity must be overcome by the momentum and mass of the water droplets from operating sprinklers.

Flue spaces are critical for the control or suppression of a fire involving commodities in fixed storage racks. Flue spaces are the open spaces between the commodities and the structural columns and beams of the rack—they are defined based on their orientation. The flue spaces that are perpendicular to the direction of loading the tiers with storage arrays are defined as longitudinal flue spaces. Longitudinal flue spaces traverse the entire length of the storage racks. Flue spaces between each rack upright column which parallel the direction of loading are transverse flue spaces. (See Figure 14-15) Flue spaces are essential in storage racks for high-piled combustible storage, because they permit ceiling sprinkler water to penetrate into commodities to either control or suppress the fire, depending on the design of the automatic sprinkler systems. **[Ref. 2308.3]**

The IFC has requirements for automated storage. While the term "automated storage" is not defined, many fire code officials correctly assume it is carousel storage. Carousel storage and retrieval systems are factory built motorized storage systems that revolve around a fixed base. In most cases, the path of the revolution has two long parallel sides connected by round, short radius ends. They use fixed tracks with the motor mounted either on the top or bottom. A conventional carousel storage system revolves in the horizontal plane, they may be vertical, or a combination of both. (See Figure 14-16) Products stored in the carousel are brought to a stationary picking station using manual or computer control. Computers are commonly used to simultaneously maintain inventory records and conduct ordering. Typically, the commodities are small parts/products. **[Ref. 2309.1]**

Automatic sprinkler protection for carousels is required when specified by Table 2306.2. Sprinkler protection for these storage systems are based on:

- Classification of the stored commodities
- Type of storage containers
- Height of storage

For horizontal carousels, NFPA 13 may require the installation of an in-rack automatic sprinkler system, especially if the height of the carousel is over 8 feet and combustible containers are used. **[Ref. 2309.2]**

Because carousels rotate on a fixed track, a significant concern is a fire involving the carousel system moving the fire throughout an area. To limit the potential for such an incident, automatic shutdown features are prescribed. This can include the installation of an automatic smoke detection system or a dead-man switch, which only allows the carousel to operate when an operator is in attendance. **[Ref. 2309.3]**

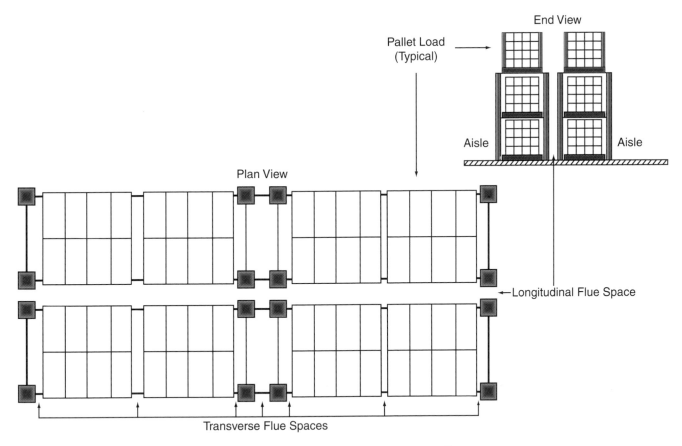

FIGURE 14-15 Longitudinal and transverse flue spaces

FIGURE 14-16 Horizontal carousel storage system

AISLES

The requirements for aisles are applicable when the high-piled combustible storage area is greater than 500 ft.² Aisles are provided to facilitate material handling and provide access for fire suppression and overhaul activities. The requirements for the minimum width of the aisles are based on if the building is or is not protected by an automatic sprinkler system. Aisles are required to comply with the requirements of NFPA 13 based on the method of storage. When applying the requirements of NFPA 13, an important variable is the minimum aisle width between storage piles or racks. Aisles are required by IFC Chapter 10 to facilitate means of egress from a building. [**Ref. 2306.9**]

A primary role of aisles is to limit the spread of fire due to radiant heat transfer. Aisles reduce the potential for fire spread from one storage pile or rack to another by providing a defined space between these areas.

Radiation heat transfer is the result of electromagnetic radiation that arises due to the temperature of a source. Because it is an electromagnetic source of radiation, heat transfer occurs without the need for air or metal. A classic example is radiant heat transfer is the sunlight of day. When a cloud covers the sun, both its heat and light diminish. Aisles provide a mechanism to reduce the potential for a fire to transfer enough energy to ignite adjacent storage piles or racks. At a minimum, NFPA 13 requires rack storage to have 4-foot aisles. As the width of aisles increases, the required sprinkler discharge is reduced because the potential for radiant heat transfer is further reduced. A basic rule of radiation heat transfer is that for every 2 feet of reduction in aisle width, the amount of radiant heat transfer is doubled when evaluating a fire in a rack or storage pile and its ability to spread to an adjacent rack by crossing over the aisle. This rule of radiation heat transfer serves as an important reason for verifying the width of aisles, especially when inspecting buildings housing high-piled combustible storage. (See Figure 14-17)

Buildings that are not protected by an automatic sprinkler system must have a minimum width of 8-feet (or 96-inch) aisles. Buildings that are equipped with sprinklers may be allowed to have aisle widths of less than 8 feet. However, this is dependent on the design of the automatic sprinkler system, the stored commodity, the storage method, and if the building is accessible to the public. The IFC allows aisles limited to employee access to be as little as 24 inches in width. [**Ref. 2306.9.1.2**]

Removing or inserting products on racks (stocking) may be done manually or mechanically. Manual stocking employs ladders or other nonmechanical equipment to move stock, while mechanical stocking uses motorized vehicles such as fork lift trucks and pallet trucks to move stock. Requirements are specified for the maintenance of aisles that stipulate the minimum aisle width required during the periods when manual or mechanical commodity stocking is underway. Aisles, exit doors, and fire department access

FIGURE 14-17 Aisles allow for movement of material handling equipment, provide an exit access path, and limit the potential for fire spread between racks

doors cannot be obstructed and shall be maintained free of obstructions. (See Figure 14-18) The IFC also prescribes that certain aisle widths be maintained during manual or mechanical stocking activities. For manual stocking operations, aisle widths must equal at least 50% of the aisle dimension for aisles wider than 48 inches. For aisles 48 inches or less in width, a minimum 24-inch aisle is required during manual stocking operations. In cases where mechanical stocking occurs, a minimum 44-inch aisle is required. [Ref. 2305.4]

FIGURE 14-18 Aisles must be maintained during stocking operations and cannot obstruct the exit access

Other Special Uses and Processes

The IFC regulates numerous hazardous and special processes and uses. Several of these activities present a potential for fires and explosions if the use or process is improperly designed, operated, or maintained. In the case of the construction and demolition of a building, these activities can compromise firefighter safety because structural components and fire protection systems are either incomplete or compromised. Activities involving the use of combustible dusts present the hazard of a deflagration which can injure and kill plant personnel. Hot work is a hazardous activity because it is a source of ignition by the nature of the activity. The special processes and uses reviewed in this chapter are combustible dust producing operations, fire safety during construction and demolition, and hot work.

COMBUSTIBLE DUST PRODUCING OPERATIONS

Combustible dust producing operations occur in a variety of industries, including food production, manufacturing of pharmaceuticals, certain wood working operations, and some plastic manufacturing processes. IFC Chapter 13 addresses the prevention of dust explosions, which technically are dust deflagrations. A deflagration is an exothermic reaction (meaning it releases heat) resulting from a rapid oxidation of a combustible dust, in which the reaction progresses through the unburned material at a rate less than the velocity of sound. Deflagrations are commonly termed as "slow explosions." Deflagrations are far more common than explosions, which have burning rates greater than the speed of sound. [Ref. 2702.1]

To produce a dust deflagration requires a combustible dust as a source of fuel. A combustible dust is *a finely divided solid material which is 420 microns or less in diameter. When dispersed in air in the proper proportions, it can be ignited by a source of ignition. Combustible dust will pass through a U.S. number 40 sieve.* [Ref. 1302.1] Particle size is important. When particles become smaller, their mass is reduced, causing their surface to become proportionally larger. This increases the potential energy in a dust deflagration because the material is more easily ignited.

A dust is combustible when its particles will burn. Table salt is small enough to be considered a dust, but it is chemically noncombustible. Wood is combustible, but dimensioned lumber is not dust: it is a solid mass with a surface area large enough to make it difficult to ignite in air using an oxygen-acetylene torch.

The energy required to ignite a combustible dust is defined as the minimum ignition energy and is measured in millijoules. A typical spark created by walking across a carpeted floor and touching a metal door is about 100 millijoules. The lower the minimum ignition energy value the less energy is required for ignition to occur. Consider, for example, an agricultural dust such as wheat flour with an average particle size of 80 microns. The minimum ignition energy required to ignite such dust is approximately 95 millijoules. If the particle size is doubled to 160 microns, the required ignition energy is over 400 millijoules. Table 15-1 lists the particle size of common materials.

TABLE 15-1 Particle sizes of common materials

Material	Size (micron)
White granulated sugar	450–600
Sand	50 and greater
Talcum powder	10
Mold spores	10–30
Human hair	40–300
Wheat, corn, or soybean flour	1–100

The mechanism of a dust deflagration requires more conditions when compared to the conventional fire triangle. Along with an ignition source and proper mixing of the fuel with an oxidizer, a dust deflagration also requires the fuel be in a confined enclosure, such as a building or exhaust duct, and that it be easily dispersed within the enclosure. A deflagration can occur when enough dust particles are suspended in the enclosure; the concentration exceeds the minimum explosive concentration and the ignition source is greater than the minimum ignition energy. The burning combustible dust liberates flammable gases and it is the ignition of these gases that causes the deflagration. (See Figure 15-1)

The IFC specifies requirements for controlling sources of ignition and housekeeping to reduce the potential of a dust deflagration. The IFC adopts NFPA standards that regulate dust deflagration hazards. To prevent the ignition of the dust layer, the IFC requires strict control of ignition sources. A means of collecting the dust is required to prevent enough of a dust accumulation that can be suspended in air. The key is to ensure the dust does not accumulate at any locations where it can be suspended in air, such as on ventilation ducts, light fixtures, building trusses or purlins, cable trays, or similar locations. (See Figure 15-2) [**Ref. 1303.1 and 1303.2**]

In two of the referenced NFPA standards adopted by the IFC, one means of determining if a dust deflagration hazard exists is to measure the dust layer depth. NFPA 654, *Manufacturing, Processing and Handling of Combustible Particulate Solids* and NFPA 664, *Prevention of Fires and Explosions in Wood Processing and Woodworking Facilities* indicate that such a hazard exists when the dust is combustible, it has a density of 75 pounds/cubic foot or less, and the area of dust layer and its depth exceed the values indicated in Table 15-2.

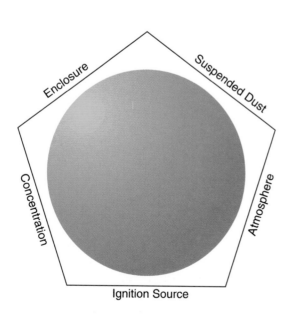

FIGURE 15-1 Dust deflagration pentagon

FIGURE 15-2 Dust collector used to capture combustible dust produced during the manufacturing of furniture

TABLE 15-2 Minimum dust layer depth and area for a dust deflagration hazard by material or facility type

Material or facility type	Dust layer depth	Dust layer area
Combustible particulate solids	1/32 inch	5% when the building area is less than 20,000 square feet, or a maximum dust layer area of 1,000 square feet when the building area is 20,000 square feet or greater.
Wood processing and woodworking facilities	1/8 inch	

FIRE SAFETY DURING CONSTRUCTION AND DEMOLITION

Erecting or demolishing a building introduces a variety of hazards. When constructing a building, unprotected shafts may be created that form a vertical path for fire travel. A variety of hazardous materials can be on site such as paints, coatings, and fuels which can contribute to fire growth and spread. When a building is demolished, structural components may be weakened or removed, which reduce the building's resistance to external loads like rain, wind, snow, or ground motion from an earthquake. Previously functioning fire protection systems (sprinklers and standpipes) may be disconnected or otherwise inoperable. IFC Chapter 14 addresses the fire safety aspects of structures being constructed, altered, or demolished. These requirements address temporary heating, establishing precautions against fire, the storage and use of hazardous materials, and maintaining fire protection systems. [Ref. 1401.1]

Temporary heating is commonly employed in buildings during construction or demolition to maintain water-based fire protection systems above water's freezing temperature and the safety of personnel. Heaters used to heat the building are required to be listed and installed with the *International Mechanical Code* (IMC) and the *International Fuel Gas Code* (IFGC). Depending on the fuel, refueling must be in accordance with applicable IFC requirements. Clearances between the heater and combustibles must be maintained to avoid ignition of construction materials. Heaters must be supervised by competent personnel. (See Figure 15-3) [Ref. 1403]

Another consideration during building construction or demolition is the type and amount of fuels available for equipment and machinery. The predominance of equipment found at these sites uses include number 2 diesel fuel, unleaded gasoline, or LP-Gas. The requirements in IFC Chapter 14 require compliance with the IFC flammable and combustible liquid and LP-Gas provisions. Any operations that require the application of flammable and combustible liquids, such as the application of paints and coatings, require these activities be performed in areas with

FIGURE 15-3 Heaters must be installed and maintained based on their available fuel and the requirements in the IMC and IFGC

FIGURE 15-4 Interior application of flammable finishes in buildings under construction or renovations are a regulated activity

FIGURE 15-5 Combination standpipe and sprinkler fire department connection

adequate ventilation. (See Figure 15-4) Note that specific requirements for such activities are also found in IFC Chapter 15 for floor finishing. **[Ref. 1405 and 1509]**

A major consideration during building construction, alternation, and demolition is ensuring the contractor and subcontractors are properly trained, aware of the fire hazards found on a construction site, and know what actions are required in the event that an unwanted fire occurs. The responsibility for implementing a fire prevention plan rests with the project superintendant. This person (or their designee) is responsible for the maintenance of any required fire protection systems and is the impairment coordinator. This person is responsible for training job-site personnel about the building's fire protection features and systems and how they are maintained and serviced. An approved preplan must be prepared for firefighters to indicate the location and types of various hazards or pitfalls the building may present. The person assigned as the fire safety superintendant is responsible for the implementation of the hot work permit program (see the next section). **[Ref. 1408]**

Maintaining fire protection systems and features is an important element of the construction, alteration, or demolition of a building. Buildings under construction or demolition involved in a fire have the greatest potential for collapse or other failures because the life safety systems are not completed or are being demolished. The IFC requires standpipe systems complying with NFPA 14, *Standard for the Installation of Standpipes and Hose Systems* be provided when a building height is over 40-feet. The standpipe system must be extended as to within one floor of the highest level of construction where secured floors or decking is present. During demolition, the standpipe system is required to be maintained within one floor being demolished. Similar requirements exist for automatic sprinkler systems. (See Figure 15-5) **[Ref. 1413]**

WELDING AND OTHER HOT WORK

The construction of buildings, machinery, appliances, and industrial processes commonly requires the use of welding or other hot work processes to coalesce various ferrous and non-ferrous metals. (See Figure 15-6) Hot work using oxygen and a fuel gas are performed to cut these metals into shapes or objects. All of these and other activities, including the installation of torch-applied roofing systems or thawing of frozen water pipes, constitute hot work and is regulated by the IFC. **[Ref. 2602.1]**

The IFC regulates hot work activities based on the type of operation performed, such as electric arc welding, gas welding and cutting, or a torch-applied roof system. These operations must satisfy the general and fire safety requirements in IFC Chapter 26. Under the IFC, an operational hot work permit is issued when the permit applicant presents an acceptable plan for managing this activity. The permitted party is then responsible for all hot work operations within their facility or building for the life of the operating permit. For certain activities such as hot work causing the impairment of an automatic sprinkler system, the potential exists for the ignition of flammable vapors or combustible materials or on board of ships at dock, specific approval by the fire code official is required. [Ref. 2601.3]

All hot work operations must be under the supervision of a responsible person. The individual is responsible for reviewing the site prior to issuing a permit and performs subsequent inspections as work progresses to ensure it is in accordance with the hot work permit and program. This includes maintaining adequate documentation demonstrating compliance with IFC Chapter 26, the hot work plan and ensuring that adequate signs are posted in the work area. (See Figure 15-7)

An area review and inspection is required prior to commencing hot work. This inspection, performed by the responsible person, ensures combustibles and building openings are properly shielded and that the area is clean of any combustible debris. When welding partitions are used, the inspection should verify the partitions prevent the passage of sparks, slag, and heat from the hot work area. In areas protected by an automatic sprinkler system the sprinklers in the immediate area can be shielded—however, these shields must be removed at the end of the work day or assignment. [Ref. 2604.1 and 2604.3]

A fire watch is required during and after hot work activities. (See Figure 15-8) The fire watch is maintained for at least 30 minutes after hot work has concluded. The individual is responsible for extinguishing any spot fires and communicating an alarm to the fire department. Fire watch personnel must be trained in the use of portable fire extinguishers. A fire watch is not required in areas with no fire hazards or no combustible materials. [Ref. 2604.2]

FIGURE 15-6 Hot work such as welding of this structural member is regulated by the IFC

FIGURE 15-7 Sign prohibiting hot work in a flammable vapor area

FIGURE 15-8 Fire watch during hot work operations

FIGURE 15-9 Metal cutting using oxygen-acetylene fuel gas

FIGURE 15-10 Oxygen-acetylene fuel gas cutting torch

The use of oxygen and a flammable gas in many cutting, welding and brazing activities. Common fuel gases include acetylene, methylacetylene-propadiene (MAPP gas), or LP-Gas. (See Figure 15-9) Of these three gases, acetylene is the most commonly used because of the high temperatures it can develop when it is properly mixed with oxygen. Oxygen-acetylene fuel gas mixtures can produce temperatures from 5,800 to 6,300°F depending on the mixture. These high temperatures can cut through a variety of metals. Acetylene is classified as compressed flammable and class 2 unstable (reactive) gas. The gas is dissolved in solution with acetone to facilitate its safe storage and handling.

An oxygen-acetylene torch is assembled using two compressed gas cylinders. (See Figure 15-10) One cylinder contains compressed or cryogenic oxygen and the other cylinder contains the fuel gas. Because the gases are stored at high pressures, a pressure regulator is installed on each cylinder to control the pressure at the torch. Downstream of the regulator discharge fitting, a hose is connected which is terminated at the cutting torch. While oxygen and all of the fuel gases are incompatible hazardous materials (see Chapter 16), the IFC allows the cylinders to be located adjacent to each other because they are equipped with pressure regulators and the low loss history involving oxygen-fuel gas cutting and welding. [Ref. 2605.2.1]

The appropriate pressure regulator is required for oxygen and the selected fuel gas. Acetylene regulators are designed to limit the discharge pressure to 15 PSIG or less. (See Figure 15-11) Acetylene becomes explosively reactive above this pressure so the IFC code establishes a 15 PSIG pressure limit. Oxygen regulators must be designed and handled so they never come into contact with any flammable or combustible liquids, including lubricants such as oil. Oil and oxygen are incompatible hazardous materials, and if oil is ever introduced into an oxygen system, the pressure regulator can explode. [Ref. 2605.3 and 2605.4]

FIGURE 15-11 Acetylene pressure regulator (*Courtesy of Smith Equipment, Watertown, SD*)

FIGURE 15-12 Installation of a torch applied roofing system (*Courtesy of Midwest Roofing Contractors Association, Lawrence KS*)

Another common hot work activity is the installation of torch-applied roofing systems. (See Figure 15-12) The roofing materials are made water tight by fusing and melting bitumen and asphalt in the roofing material to the roof's substrate. The fusing of these materials is commonly performed using specialized LP-Gas fueled burners. Torch applied roofing systems should only be performed by individuals who have been trained in the fire safety requirements for the application of these materials, the safe storage and handling of LP-Gas, and in the use of portable fire extinguishers. **[Ref. 1417.1]**

PART VI

Hazardous Materials

Chapter 16: General Requirements for Hazardous Materials

Chapter 17: Compressed Gases

Chapter 18: Flammable and Combustible Liquids

CHAPTER 6

General Requirements For Hazardous Materials

Requirements for the storage, handling, use, and dispensing of hazardous materials represents the largest body of regulations in the IFC. Hazardous materials can be a challenge because their proper classification is essential in determining and applying the IFC and IBC requirements. In some cases, determining the proper classification itself can be a significant challenge that requires technical assistance from a variety of qualified and competent sources. Not all chemicals and chemical compounds are hazardous materials as defined by the IFC. Chemicals and chemical formulations that do not fall into one of the 12 IFC hazard categories are exempt from regulation under the IFC and IBC. Currently, the American Chemistry Society has over 1 million chemicals or formulations registered in the Chemical Abstract Service. The IFC regulates only about five percent of these chemicals—however, the chemicals that are regulated are essential in the manufacturing of numerous consumer and industrial products.

Chapter 16 reviews the general requirements for hazardous materials. IFC Chapter 27, *General Requirements for Hazardous Materials* is applied in conjunction with the material-specific requirements in IFC Chapters 28 through 44. Chapters 17 and 18 of this text focus on the two forms of hazardous materials that are encountered by all fire code officials: compressed gases and flammable and combustible liquids.

The requirements in IFC Chapter 27 are used in conjunction with the specific regulations for each class of hazardous materials regulated. For example, consider a building that is storing 300 gallons of a corrosive liquid. Proper code application would require the enforcement of the requirements in IFC Chapters 27 and 31—Chapter 27 for the general hazardous material provisions and Chapter 31 for the corrosive material provisions. Chapter 27 contains definitions and general requirements in Sections 2702 and 2703 that are applicable regardless of the quantity of hazardous materials in storage or use. Section 2704 contains requirements that are applicable when the amount of storage exceeds a certain quantity threshold, known as a Maximum Allowable Quantity per Control Area (MAQ). Section 2705 specifies requirements that are applicable when the hazardous materials are being used or dispensed, and sets forth specific provisions when the amount of hazardous materials in use exceeds the MAQ.

Chapter 27 of the IFC sets forth the minimum requirements for the storage, use, handling, and dispensing of hazardous materials. The terms *storage, use, handling,* and *dispensing* are defined in the fire code, and understanding them is important to the proper application of the code. These terms describe the "environment" in which hazardous materials occur, and they affect the MAQ. Each class of hazardous material is assigned a MAQ based on its hazards and the relative risk of the material to the building and its occupants. For example, materials in "storage" exist in their original, unopened containers such as those one might find in the paint section of a hardware store. Since the packaging has not been opened and the contents have not mixed with the atmosphere, the code allows a greater quantity to occur within a specific area than if the same product were in "use" or "dispensing" where the material could be spilled, spread, or otherwise exposed to an environment where it creates a hazard. Remember that as long as hazardous materials remain in their containers, generally they are easier to control.

The requirements in Chapter 27 apply to new (i.e., virgin or unused) and waste chemicals, including their transportation on the site where they are stored, used, handled, or dispensed.

Certain processes are exempt from the Chapter 27 requirements. Exemption from these requirements does not relieve the permit applicant from compliance with other IFC provisions and the other adopted codes in the jurisdiction. Exemption also does not relieve the permit applicant from the IFC hazardous materials classification requirements. Off-site transportation is regulated by the U.S. Department of Transportation and is exempt from the IFC regulations. (Review Chapter 1 for a review of off-site transportation requirements.) (See Figure 16-1) **[Ref. 2701.1, exception 3]**

The IFC exemptions address specific processes or activities that store or use hazardous materials. Some of the exemptions concern activities that are regulated by other ICC code provisions and others exempt activities regulated and pre-empted by Federal law. For example, mechanical refrigeration systems, which can use refrigerants that are flammable or are corrosive (such as anhydrous ammonia) are exempt from the Chapter 27 requirements but are regulated by the requirements in Section 606 and the *International Mechanical Code.* Stationary storage battery systems using corrosive materials suspended in an aqueous gel are regulated by the requirements in Section 609. These specific provisions offer greater safety when compared to the general and material specific requirements in Chapters 27 through 44. [**Ref. 2701.1, exceptions**]

MATERIAL CLASSIFICATION

Before one can apply many of the IFC requirements the building's occupancy classification must be assigned. A similar exercise is required when applying the IFC regulations for the storage, handling, dispensing and use of hazardous materials: they must be classified based on their physical and health hazards, physical state, and environment. The IFC classification system recognizes that many hazardous materials represent more than one hazard—all of the hazards of the particular chemical or compound must be addressed. Each hazardous material subject to regulation must be evaluated and assigned its hazard classifications and ratings based on all of its hazards. [**Ref. 2701.2**]

FIGURE 16-1 This natural gas pipeline is regulated by the U.S. Department of Transportation – Office of Pipeline Safety and Hazardous Materials. The Federal Pipeline Safety Act preempts the IFC requirements

Hazardous materials classification can be challenging. Classifying hazardous materials requires a basic understanding of each class of hazardous material regulated by the IFC. Classifications can be complicated when a material is diluted in water, intentionally mixed with other chemicals, or its physical form or state is modified. The IFC allows classifications to be assigned using nationally recognized standards, an approved organization or individual, Material Safety Data Sheets (MSDS) or by other approved means. [Ref. 2701.2.1]

A number of sources are available to assist in classification of hazardous materials. IFC Appendix E contains examples of selected hazardous materials based on the Chapter 27 classification system. Another economical and reliable information source available from ICC is the Hazardous Materials Expert Assistant (HMEX). (See Figure 16-2) HMEX is an electronic data base cataloging over 8,000 hazardous materials, compounds, and product formulations based on the IFC classification criteria. It also contains basic physical data for each material.

Hazardous materials are commonly manufactured, shipped, and consumed as a mixture with some diluent such as water or nitrogen gas. Diluted materials fulfill a market or customer requirement. In other cases chemicals are not diluted but may be mixed together to serve a particular purpose or application. Mixtures are required to be classified based on their hazards as a whole. [Ref. 2701.2.1]

Consider a mixture of sulfuric acid and water. Sulfuric acid is used in the manufacturing of aerospace, automotive, food, electronics, textile, pharmaceutical, and plastics products. Like other hazardous materials, its hazards can change based on the amount (sometimes termed as "strength") of sulfuric acid in solution with water. As illustrated in Table 16-1, the hazards of sulfuric acid change when less acid is added to water.

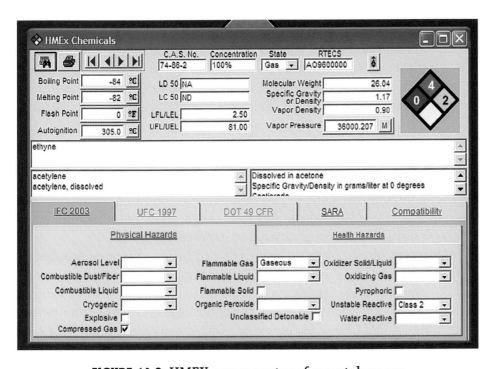

FIGURE 16-2 HMEX screen capture for acetylene gas

TABLE 16-1 Hazard classifications of selected concentrations of sulfuric acid

Concentration (% volume of sulfuric acid in water)	IFC Hazard Classification	Common uses
98%	Class 2 Water Reactive, Corrosive, Toxic Liquid	Fertilizer manufacturing; cleaning of semiconductor wafers
32–38%	Class 1 Water Reactive, Corrosive Liquid	Electrolyte in wet-cell automobile batteries
3%	Nonhazardous	Cutting an onion releases an amino acid known as a sulfoxide. When the sulfoxide mixes with the water in a human eye, it forms a weak solution of sulfuric acid.

The IFC hazardous materials classification system is based on the classification system used by the U.S. Department of Labor, Occupational Safety and Health Administration (OSHA). The OSHA criterion is used because its classification method focuses on workplace safety, which matches the intent and scope of the fire code. The IFC classification system is generally not consistent with the U.S. DOT or Environmental Protection Agency classification systems. These and other Federal agencies have different classification criterion because of the differences in and the goals of their regulations. [Ref. 2701.2.2]

Hazardous materials classified using the IFC scheme are physical hazards, health hazards, or a combination of both. A single hazardous material can present one or more physical and health hazards. Physical hazard materials burn, accelerate burning, and may either detonate or deflagrate. Many of the physical hazard materials that burn can produce a deflagration under certain conditions, such as when heated or if they are pressurized. Conversely, certain classes of oxidizers and most classes of explosives are capable of producing detonations. Oxidizers are the only physical hazard that can chemically accelerate burning. [Ref. 2701.2.2.1]

Many physical hazard materials are assigned a class designation. The class designation is a relative estimation of the potential outcome if the hazardous material is improperly stored or used. The class designation provides an indication of the ease or difficulty of igniting the material or initiating an uncontrolled chemical reaction and the potential outcomes if the material is improperly stored or handled.

Different physical hazard materials can have a different number of hazard classes. The class number convention is not consistent among all the physical hazard products. For example, a Class 4 oxidizer is more hazardous than a Class 1 oxidizer, the opposite is true for organic peroxides where a Class I organic peroxide is more dangerous than a Class IV organic peroxide. This anomaly in the classification criteria is a result of how hazardous materials classification criteria in consensus safety standards were developed before they were incorporated into the model codes. (See Figure 16-3)

IFC Chapter 27 regulates the following hazardous materials that are a physical hazard:

1. Explosives and blasting agents
2. Combustible liquids
3. Flammable solids, liquids, and gases
4. Organic peroxide
5. Oxidizer, solids, or liquids
6. Oxidizing gases
7. Pyrophoric solids, liquids, or gases
8. Unstable (reactive) solids, liquids, or gases
9. Water-reactive materials solids or liquids
10. Cryogenic fluids

Arabic Numeral	GREATEST POTENTIAL	Roman Numeral
4		I
3		II
2		III
		IV
1	LEAST POTENTIAL	V

FIGURE 16-3 Physical Hazards Numbering Convention

Code Basics

Physical hazards are capable of burning, accelerating burning, deflagrations, or detonations. Many of the physical hazard materials are assigned a class designation, which is relative estimation of the outcome if the hazardous material is not correctly stored and handled. ●

The 14th century alchemist Paracelsus, considered by some to be the father of toxicology, wrote: *[a]ll things are poison and nothing is without poison, only the dose permits something not to be poisonous.* Applying this philosophy, materials such as water and dirt are poisons, and annually people die because of drowning or being trapped when an excavated trench collapses. The IFC regulates health hazard materials that cause death, injury, or incapacitation if an individual has a single, brief exposure to the hazardous material.

The ratings for health hazard materials are based on analytical and laboratory testing of living organisms and the three routes that the human body can be exposed to hazardous materials: inhalation, absorption, and ingestion. The results of these tests determine if a health hazard material is classified as Highly Toxic, Toxic, or Corrosive. [Ref. 2701.2.2.2]

The classification of a health hazard material as being Highly Toxic or Toxic is based on reproducible results from tests supervised by toxicologists. These tests, in conjunction with other scientific literature, are used to establish the values known as LC_{50} and LD_{50} (Table 16-2). The term LC_{50} is a measurement of the lethal concentration of a hazardous material that kills 50% of the animals tested. LC_{50} values are measurements of toxins that are inhaled and the measurement is expressed in milligrams of toxin/liter of air (Mg/L) or in parts per million or billion (PPM or PPB) of the contaminant in air. LD_{50} is the measurement of the amount of a toxin that kills 50% of the test animals when exposed to the chemical via absorption through skin or by ingestion. LD_{50} values are expressed in milligrams of toxin/kilograms of animal weight (Mg/Kg). The lower the LD_{50} or LC_{50} value, the more powerful the toxin because it requires less of the material to cause 50% of the test animals to die. The IFC criterion for classifying these materials is based on specific animal species of a certain body weight. [Ref. 3702.1]

TABLE 16-2 IFC Classification Criterion for Highly Toxic and Toxic Materials

IFC Classification	Exposure Route		
	Inhalation Toxicity Threshold	Absorption Toxicity Threshold	Ingestion Toxicity Threshold
Highly Toxic	200 PPM or less; 2 Mg/L or less	200 Mg/Kg or less	50 Mg/Kg or less
Toxic	Greater than 200 PPM but not more than 2,000 PPM; greater than 2 Mg/L but not more than 20 Mg/L	Greater than 200 Mg/Kg but not more than 1,000 Mg/Kg	Greater than 50 Mg/Kg but not more than 500 Mg/Kg

Mg/Kg = milligram/kilogram of body weight

Mg/L = milligram/liter of mist, fume, or dust

PPM = Parts per million of gas or vapor

FIGURE 16-4 DOT Corrosive Label

Materials that cause visible destruction or irreversible alterations in living tissues by a chemical action are classified as corrosive. Corrosives can be a solid, liquid, or gas. The IFC definition of Corrosive specifically excludes materials that chemically react when they contact an inanimate object, such as metals. One example is ferric chloride used in water treatment and metal etching. Ferric chloride aggressively attacks most metals yet will not harm human skin. A hazardous material is classified as a corrosive based on the test method specified by the U.S. DOT. Because the IFC references the DOT test method, a hazardous material with a corrosive label or placard is generally assigned the same IFC hazard classification. (See Figure 16-4)

EXAMPLE
Acrolein is used in the manufacturing of a number of plastics. Acrolein has a flash point temperature of −15°F and a boiling point temperature of 127°F. The material has an absorption LD_{50} value of 200 Mg/Kg and an ingestion LD_{50} value of 26 Mg/Kg. Based on the definitions and classifications for flammable and combustible liquids in IFC Section 3402 and highly toxic and toxic materials in IFC Section 3702, what is its classification?

ANSWER
Acrolein is a Class I-B Flammable and Highly Toxic liquid.

HAZARDOUS MATERIALS REPORTING

The requirement for an IFC operational permit to store, handle, use, and dispense a hazardous material is based on the permit applicant reporting the hazardous material classification, physical state, the method it will be used, and the amount of material. When required by the fire code official, the applicant for a hazardous materials construction or operational permit may be required to submit either a Hazardous Materials Management Plan (HMMP) or a Hazardous Materials Inventory Statement (HMIS). (See Figure 16-5)

The HMMP documents basic information so emergency responders understand the basic construction and access routes for a building or premises, such as emergency exits that can be used for access, the physical and health hazards of hazardous materials stored and used within particular areas, where emergency responders will meet the fire department liaison, and the location of all aboveground and underground tanks, sumps, vaults, or any other below-grade processes. This information is extremely beneficial to emergency responders because it identifies locations where confined space entries may need to be performed. The HMMP also identifies

Code Basics

LC_{50} values measure the toxic effects of hazardous materials whose route of exposure is inhalation. LD_{50} values measure toxic effects of hazardous materials whose route of exposure is through skin absorption or ingestion. ●

HAZARDOUS MATERIALS MANAGEMENT PLAN
SECTION I: FACILITY DESCRIPTION

1. Business Name: _____ Phone: Address:
2. Person Responsible for the Business Name   Title   Phone

_____ _____   _____   _____

3. Emergency Contacts: Name   Title   Home Number   Work Number

_____   _____   _____   _____

_____   _____   _____   _____

_____   _____   _____   _____

_____   _____   _____   _____

4. Person Responsible for the Application/Principal Contact: Name   Title   Phone

_____   _____   _____

5. Principal Business Activity:

_____

_____

_____

6. Number of Employees:
7. Number of Shifts: _____
   a. Number of Employees per Shift

_____

_____

_____

8. Hours of Operation:

FIGURE 16-5 Hazardous Materials Management Plan cover page

FIGURE 16-6 The fire department is authorized to recover the costs associated with the response to this unauthorized discharge and subsequent flammable liquid fire

building areas constructed as control areas[1] and Group H occupancies. [**Ref. 2701.5.1**]

A HMIS contains information beneficial to building and fire code officials, plans examiners, and inspectors attempting to determine that the amount of hazardous materials in storage and use complies with the IFC and IBC requirements. An HMIS documents the product's name, its chemical constituents along with their respective Chemical Abstract Service (CAS) number, the volume of containers or tanks, the product's hazard classification, and the amount in storage, use-open, and use-closed systems. This information is beneficial for plans examiners to confirm the occupancy classification, if the mechanical ventilation system should comply with the *International Mechanical Code* requirements for hazardous exhaust systems, and to verify the design of the process(es) complies with the applicable requirements of the IFC. [**Ref. 2701.5.2**]

The IFC generally requires any accidental hazardous material release be reported to the fire code official. *A hazardous material release in a manner that does not conform to the provisions of the IFC or applicable health and safety regulation is* defined as an Unauthorized Discharge. When an unauthorized discharge of hazardous materials results from a container failure, the fire code official is authorized to require its repair or removal from service. Those responsible for the unauthorized discharge are responsible for the costs associated to cleanup any hazardous material release. If fire department resources such as protective clothing, fire-fighting foam, or personnel costs are expended to respond and mitigate a unauthorized discharge, the fire code official is authorized to recover the costs. (See Figure 16-6) [**Ref. 2702.1, 2703.3.1**]

STORAGE AND USE

The situation of a hazardous material is an important part of the IFC and IBC hazardous materials regulations. The IFC recognizes hazardous materials can be in one of three situations:

- Storage
- Use – Closed System
- Use – Open System

Storage is defined as *the keeping, retention, or leaving of hazardous materials in closed containers, tanks, cylinders, or similar vessels; or vessels*

---

[1]A "control area" is a space within the building where a limited quantity (defined in Section 2702.1 as the MAQ or "Maximum Allowable Quantity per Control Area") of hazardous materials can be stored or used without affecting the occupancy classification. This concept is explained in more detail later in this chapter.

supplying operations through closed connections to the vessel. A stored hazardous material is essentially static and dormant and are generally awaiting use. IFC Section 2704 requirements that are applicable to storage exceeding the fire code quantity limit, which are based on the material's classification and physical state. [Ref. 2702.1]

A material that is "placed into action" is defined by the IFC as being in "use". Use occurs when a material's stored energy is released, kinetic energy is introduced, mixtures are created, or gravity is used to facilitate its movement. IFC Section 2705 has requirements that are applicable to use that exceeds the fire code quantity limit, which are based on the material's classification and physical state. The use of hazardous materials is regulated because placing the material into action introduces the possibility of an unauthorized discharge and therefore the piping or process systems must be reviewed and approved by the fire code official. Use can be within a closed or open system.

A closed use system is the *use of a solid or liquid hazardous material involving a closed vessel or system that remains closed during normal operations where vapors emitted by the product are not liberated outside of the vessel or system and the product is not exposed to the atmosphere during normal operations; and all uses of compressed gases.* Examples of closed systems for solids and liquids include product conveyed through a piping system into a closed vessel, system, or piece of equipment. [Ref. 2702.1]

A open use system involves solid or liquid hazardous materials. It does not include compressed gases. Compressed gases are not used in an open system. A use-open system is *the use of a solid or liquid hazardous material involving a vessel or system that is continuously open to the atmosphere during normal operations and where vapors are liberated, or the product is exposed to the atmosphere during normal operations.* Open systems for solids and liquids include dispensing from or into open beakers or containers, dip tanks, and plating tank operations. (See Figure 16-7) [Ref. 2702.1]

MAXIMUM ALLOWABLE QUANTITY PER CONTROL AREA

The amount of hazardous material allowed to be stored and used inside of a building is based on the material's hazard classification, situation, and physical state. The maximum allowable quantity (MAQ) is exactly as the term implies: it is the maximum amount of a class of hazardous material that is permitted within a building without requiring the building to be classified as a Group H (hazardous) occupancy. The IFC MAQ values were developed based on the hazards associated with each regulated class of hazardous materials.

The MAQ values in the IFC are found in Table 16-3 (IFC Tables 2703.1.1(1)) through (4). (See Table 16-3) Table 2703.1.1(1) sets forth the MAQ for physical hazard materials inside of buildings, and Table 2703.1.1(2) sets forth the health hazard materials MAQ inside of buildings. There are four MAQ tables in the IFC - two for indoor storage and use of hazardous materials and two for out storage and use. (In the

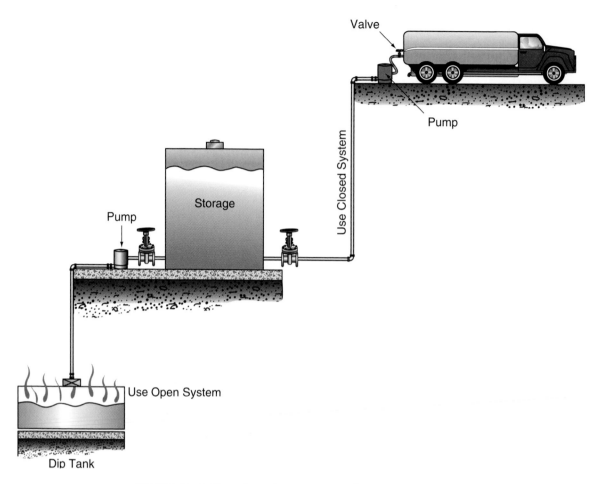

FIGURE 16-7 Use-closed, storage, and use-open systems

context of this book, Table 2703.1.1(1) is provided to explain the concepts of MAQ and control areas. Readers are encouraged to review the 2009 IFC to understand and apply the concept of MAQ and control areas for health hazard materials and outdoor control areas).

Notice that the tabular headings include the description "Maximum Allowable Quantity *per Control Area*...[Emphasis added.]" This is an important distinction for the code official and the regulated community. A control area is a space inside or outside of a building where the quantity of hazardous materials stored, used, handled, or dispensed does not exceed the MAQ. With the addition of two or more control areas, an occupant can increase the amount of hazardous materials in a building without having to reclassify it as a Group H occupancy.

The IFC MAQ tables are divided into five major headings:

- The classification of the hazardous material
- The occupancy classification of the building when the MAQ is exceeded
- Storage
- Use-Closed System
- Use-Open System [Table 2703.1.1(1) (See Table 16-3)]

Under the headings for storage and use, each group of table columns is divided by the three physical states and the unit of measurement.

TABLE 16-3 Maximum Allowable Quantity Per Control Area of Hazardous Materials Posing a Physical Hazard[a, j, m, n, p] (IFC Table 2703.1.1(1))

Material	Class	Group When the Maximum Allowable Quantity Is Exceeded	Storage[b]		Use-Closed Systems[b]			Use-Open Systems[b]		
			Solid pounds (cubic feet)	Liquid gallons (pounds)	Gas cubic feet at NTP	Solid pounds (cubic feet)	Liquid gallons (pounds)	Gas cubic feet at NTP	Solid pounds (cubic feet)	Liquid gallons (pounds)
Combustible liquid[c, i]	II	H-2 or H-3	Not Applicable	120[d, e]	Not Applicable	Not Applicable	120[d]	Not Applicable	Not Applicable	30[d]
	IIIA	H-2 or H-3		330[d, e]			330[d]			80[d]
	IIIB	Not Applicable		13,200[e, f]			13,200[f]			3,300[f]
Combustible fiber	Loose	H-3	(100)	Not Applicable	Not Applicable	(100)	Not Applicable	Not Applicable	(20)	Not Applicable
	Baled[o]		(1,000)			(1,000)			(200)	
Cryogenic Flammable	Not Applicable	H-2	Not Applicable	45[d]	Not Applicable	Not Applicable	45[d]	Not Applicable	Not Applicable	10[d]
Consumer fireworks (Class C Common)	1.4G	H-3	125[d, e, l]	Not Applicable	Not Applicable	Not Applicable	Not Applicable	Not Applicable	Not Applicable	Not Applicable
Cryogenic Oxidizing	Not Applicable	H-3	Not Applicable	45[d]	Not Applicable	Not Applicable	45[d]	Not Applicable	Not Applicable	10[d]
Explosives	Division 1.1	H-1	1[e, g]	(1)[e, g]	Not Applicable	0.25[g]	(0.25)[g]	Not Applicable	0.25[g]	(0.25)[g]
	Division 1.2	H-1	1[e, g]	(1)[e, g]		0.25[g]	(0.25)[g]		0.25[g]	(0.25)[g]
	Division 1.3	H-1 or H-2	5[e, g]	(5)[e, g]		1[g]	(1)[g]		1[g]	(1)[g]
	Division 1.4	H-3	50[e, g]	(50)[e, g]		50[g]	(50)[g]		50[g]	(50)[g]
	Division 1.4G	H-3	125[d, e, l]	Not Applicable		0.25[g]	0.25[g]		0.25[g]	0.25[g]
	Division 1.5	H-1	1[e, g]	(1)[e, g]		0.25[g]	(0.25)[g]		0.25[g]	(0.25)[g]
	Division 1.6	H-1	1[d, e, g]	Not Applicable		Not Applicable	Not Applicable		Not Applicable	Not Applicable

TABLE 16-3 Maximum Allowable Quantity Per Control Area of Hazardous Materials Posing a Physical Hazard[a, j, m, n, p] (IFC Table 2703.1.1(1)) —Cont'd

Flammable gas	Gaseous Liquefied	H-2	Not Applicable	Not Applicable	1,000[d, e]	Not Applicable	Not Applicable	1,000[d, e]	Not Applicable	Not Applicable
				(150)[d, e]	Not Applicable		(150)[d, e]	Not Applicable		
Flammable liquid[c]	IA	H-2 or H-3	Not Applicable	30[d, e]	Not Applicable	Not Applicable	30[d]	Not Applicable	Not Applicable	10[d]
	IB and IC			120[d, e]			120[d]			30[d]
Flammable liquid, combination (IA, IB, IC)	Not Applicable	H-2 or H-3	Not Applicable	120[d, e, h]	Not Applicable	Not Applicable	120[d, h]	Not Applicable	Not Applicable	30[d, h]
Flammable solid	Not Applicable	H-3	125[d, e]	Not Applicable	Not Applicable	125[d]	Not Applicable	Not Applicable	25[d]	Not Applicable
Inert Gas	Gaseous	Not Applicable	Not Applicable	Not Applicable	Not Limited	Not Applicable	Not Applicable	Not Limited	Not Applicable	Not Applicable
	Liquefied	Not Applicable	Not Applicable	Not Applicable	Not Limited	Not Applicable	Not Applicable	Not Limited	Not Applicable	Not Applicable
Cryogenic Inert	Not Applicable	Not Applicable	Not Applicable	Not Applicable	Not Limited	Not Applicable	Not Applicable	Not Limited	Not Applicable	Not Applicable
Organic peroxide	UD	H-1	1[e, g]	(1)[e, g]	Not Applicable	0.25[g]	(0.25)[g]	Not Applicable	0.25[g]	(0.25)[g]
	I	H-2	5[d, e]	(5)[d, e]		1[d]	(1)[d]		1[d]	(1)[d]
	II	H-3	50[d, e]	(50)[d, e]		50[d]	(50)[d]		10[d]	(10)[d]
	III	H-3	125[d, e]	(125)[d, e]		125[d]	(125)[d]		25[d]	(25)[d]
	IV	Not applicable	Not Limited	Not Limited		Not Limited	Not Limited		Not Limited	Not Limited
	V	Not applicable	Not Limited	Not Limited		Not Limited	Not Limited		Not Limited	Not Limited
Oxidizer	4	H-1	1[g]	(1)[e, g]	Not Applicable	0.25[g]	(0.25)[g]	Not Applicable	0.25[g]	(0.25)[g]
	3[k]	H-2 or H-3	10[d, e]	(10)[d, e]		2[d]	(2)[d]		2[d]	(2)[d]
	2	H-3	250[d, e]	(250)[d, e]		250[d]	(250)[d]		50[d]	(50)[d]
	1	Not Applicable	4,000[e,f]	(4,000)[e,f]		4,000[f]	(4,000)[f]		1,000[f]	(1,000)[f]

Oxidizing gas	Gaseous	H-3	1,500[d,e]	Not Applicable	Not Applicable	1,500[d,e]	Not Applicable	Not Applicable	Not Applicable	Not Applicable
	Liquefied		Not Applicable	(150)[d,e]	Not Applicable	Not Applicable	(150)[d,e]	Not Applicable	Not Applicable	Not Applicable
Pyrophoric	Not Applicable	H-2	50[e,g]	(4)[e,g]	4[e,g]	10[e,g]	(1)[g]	1[g]	0	0
Unstable (reactive)	4	H-1	10[e,g]	(1)[e,g]	1[e,g]	2[e,g]	(0.25)[g]	0.25[g]	0.25[g]	(0.25)[g]
	3	H-1 or H-2	50[d,e]	(5)[d,e]	5[d,e]	10[d,e]	(1)[d]	1[d]	1[d]	(1)[d]
	2	H-3	250[d,e]	(50)[d,e]	50[d,e]	250[d,e]	(50)[d]	50[d]	10[d]	(10)[d]
	1	Not Applicable	Not Limited	Not Limited	Not Limited	Not Limited	Not Limited	Not Limited	Not Limited	Not Limited
Water reactive	3	H-2	Not Applicable	(5)[d,e]	5[d,e]	Not Applicable	(5)[d]	5[d]	1[d]	(1)[d]
	2	H-3	Not Applicable	(50)[d,e]	50[d,e]	Not Applicable	(50)[d]	50[d]	10[d]	(10)[d]
	1	Not Applicable	Not Limited	Not Limited	Not Limited	Not Limited	Not Limited	Not Limited	Not Limited	Not Limited

(Continued)

TABLE 16-3 Maximum Allowable Quantity Per Control Area of Hazardous Materials Posing a Physical Hazard[a, j, m, n, p] (IFC Table 2703.1.1(1)) —Cont'd

For SI: 1 cubic foot = 0.02832 m³, 1 pound = 0.454 kg, 1 gallon = 3.785 L.

a. For use of control areas, see Section 2703.8.3.
b. The aggregate quantity in use and storage shall not exceed the quantity listed for storage.
c. The quantities of alcoholic beverages in retail and wholesale sales occupancies shall not be limited providing the liquids are packaged in individual containers not exceeding 1.3 gallons. In retail and wholesale sales occupancies, the quantities of medicines, foodstuffs, consumer or industrial products, and cosmetics containing not more than 50 percent by volume of water-miscible liquids with the remainder of the solutions not being flammable shall not be limited, provided that such materials are packaged in individual containers not exceeding 1.3 gallons.
d. Maximum allowable quantities shall be increased 100 percent in buildings equipped throughout with an approved automatic sprinkler system in accordance with Section 903.3.1.1. Where Note e also applies, the increase for both notes shall be applied accumulatively.
e. Maximum allowable quantities shall be increased 100 percent when stored in approved storage cabinets, day boxes, gas cabinets, exhausted enclosures, or listed safety cans. Listed safety cans shall be in accordance with Section 2703.9.10. Where Note d also applies, the increase for both notes shall be applied accumulatively.
f. Quantities shall not be limited in a building equipped throughout with an approved automatic sprinkler system in accordance with Section 903.3.1.1.
g. Allowed only in buildings equipped throughout with an approved automatic sprinkler system.
h. Containing not more than the maximum allowable quantity per control area of Class IA, Class IB or Class IC flammable liquids.
i. The maximum allowable quantity shall not apply to fuel oil storage complying with Section 603.3.2.
j. Quantities in parenthesis indicate quantity units in parenthesis at the head of each column.
k. A maximum quantity of 200 pounds of solid or 20 gallons of liquid Class 3 oxidizers is allowed when such materials are necessary for maintenance purposes, operation or sanitation of equipment when the storage containers and the manner of storage are approved.
l. Net weight of pyrotechnic composition of the fireworks. Where the net weight of the pyrotechnic composition of the fireworks is not known, 25 percent of the gross weight of the fireworks including packaging shall be used.
m. For gallons of liquids, divide the amount in pounds by 10 in accordance with Section 2703.1.2.
n. For storage and display quantities in Group M and storage quantities in Group S occupancies complying with Section 2703.11, see Table 2703.11.1.
o. Densely-packed baled cotton that complies with the packing requirements of ISO 8115 shall not be included in this material class.
p. The following shall not be included in determining the maximum allowable quantities:
1. Liquid or gaseous fuel in fuel tanks on vehicles.
2. Liquid or gaseous fuel in fuel tanks on motorized equipment operated in accordance with this code.
3. Gaseous fuels in piping systems and fixed appliances regulated by the *International Fuel Gas Code.*
4. Liquid fuels in piping systems and fixed appliances, regulated by the *International Mechanical Code.*

Depending on the material classification, the MAQ for liquids is based on its volume (gallons) or weight (pounds) of material. Consider that a Class III organic peroxide has a MAQ of 125 pounds if it is a solid or liquid. The reason is most processes that use this particular class of hazardous material measure the quantity based on weight (pounds) rather than volume (gallons). To differentiate the correct method of measurement, MAQ values that are parenthetically cited are based on the measurement value referenced at the top of the column in the "Material" row of Tables 2703.1.1(1) (See Table 16-3) and 2703.1.1(2). In instances where a material is reported in gallons and is regulated by weight, the IFC uses a conversion factor of 10 pounds/gallon unless the liquid's density or specific gravity is provided. [Ref. 2703.1.2]

The IFC is conservative in it requirements for materials in use. For materials used in an open-system, the code establishes a much lower MAQ when compared to materials in closed systems. (See Figure 16-8) Consider a Class I-C flammable liquid. Its use-closed system MAQ is 120 gallons versus a use-open system MAQ of 30 gallons. Similar reductions are presented for other physical and health hazard materials. The reason for the reduction is a use-open system presents a greater number of hazards when compared to a use-closed system because it is constantly open to the atmosphere of a room and is liberating vapors, which increases the likelihood of the vapors being ignited or causing a chemical exposure. If multiple materials are used in open systems, this increases the potential of an accidental mixing of incompatible hazardous materials. Accordingly, the code reduces the MAQ for materials used in open-systems. [Ref. Tables 2703.1.1(1) and 2703.1.1(2)]

FIGURE 16-8 This open dip tank of a combustible liquid is an example of use-open system

When using Tables 2703.1.1(1) through (4) for determining the MAQ in one or more control areas, it is important to review and apply the various footnotes. The footnotes provide supplemental requirements for the proper application of these tables. Footnote "b" in Tables 2703.1.1(1), (3) and (4) and footnote "d" in Table 2703.1.1(2) stipulates that the aggregate quantity in storage and in use cannot exceed the MAQ listed for storage. For example, the storage MAQ for a Class III-A combustible liquid is 330 gallons and its use-open system MAQ is 80 gallons. Assume that an industrial process requires the open use of a Class III-A combustible liquid and the dip tank volume is 80 gallons. The proper application of this footnote would result in 250 gallons in storage and 80 gallons in a use-open system. (See Figure 16-9) [Ref. Table 2703.1.1(1)]

Certain hazardous materials present such a high fire or explosion risk that the IFC requires the installation of an automatic sprinkler system throughout the building. Footnote "g" in Table 2703.1.1(1) only permits the storage of explosive materials, pyrophorics, and certain classes of organic peroxides, oxidizers, and unstable (reactive) materials inside

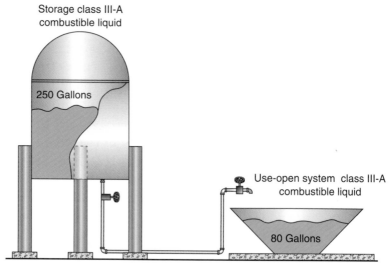

Storage class III-A combustible liquid

250 Gallons

Use-open system class III-A combustible liquid

80 Gallons

The aggregate amount in storage & use cannot exceed the total amount of storage.

FIGURE 16-9 Storage and a use-open system

FIGURE 16-10 This gas cabinet was involved in a fire resulting from a release of silane. Silane is pyrophoric and flammable gas. Because of the hazards with silane, the room and gas cabinet were required by the IFC to be protected by an approved automatic sprinkler system, which effectively controlled the fire

of buildings when they are protected by an approved automatic sprinkler system. (See Figure 16-10) The automatic sprinkler system must be installed in accordance with the requirements in Section 903.3.1.1. **[Ref. Table 2703.1.1(1), footnote "g"]**

The IFC allows the amount of hazardous materials in a control area to be increased for most physical hazard and all health hazard materials when a building is protected throughout by an approved automatic sprinkler system that complies with the requirements in Section 903.3.1.1. Table 2703.1.1(1) footnote "d" and footnote "e" in Table 2703.1.1(2) permits a 100% quantity increase when automatic sprinkler protection is provided throughout the building. (See Figure 16-11) This increase is allowed because of the reliability of an approved automatic sprinkler system to control a fire. For example, the MAQ for a solid Class 2 Oxidizer is 250 pounds in a nonsprinklered building. If the building is protected by an approved sprinkler system, the MAQ is increased to 500 pounds. **[Ref. Table 2703.1.1(1)]**

When a building is not protected by an automatic sprinkler system, the IFC allows the quantity to be increased when hazardous materials are stored in approved storage cabinets, day boxes, gas cabinets, or exhausted enclosures. (See Figure 16-12) Table 2703.1.1(1) footnote "e" and footnote "f" in Table 2703.1.1(2) allows a 100% increase of the MAQ when one of these storage methods is used. Each indicated method protects the hazardous material by providing a fire-resistive enclosure or second means of containment for the stored product. Doors to the enclosures are generally required to be self-closing and self-latching. When used in conjunction with an automatic sprinkler system, the MAQ can be increased by 400%. For example, the MAQ for a Class 2 Unstable (Reactive) liquid is 50 pounds. If the building is protected by an automatic sprinkler system complying with the requirements in Section 903.3.1.1, the MAQ increases to 100 pounds. If all of the liquid is stored in an approved hazardous materials storage cabinet, the MAQ increases to 200 pounds. **[Ref. Table 2703.1.1(1), footnotes "d" and "e"]**

FIGURE 16-11 Installation of approved automatic sprinkler system increases the MAQ by 100% for many hazardous materials

FIGURE 16-12 Storage of hazardous materials in approved cabinets is a permitted method of increasing the MAQ of many hazardous materials

CONTROL AREAS

For the purpose of regulating hazardous materials, all buildings and facilities are considered to be a control area. A control area is a space inside a building or an outdoor facility where the quantity of hazardous materials stored, used, handled, or dispensed does not exceed the MAQ. A building can have more than one control area. The number of control areas and the MAQ of hazardous materials depends on the number of stories and below grade levels in a building. [Ref. 2703.8.3]

When a building is divided into different spaces for the storage and use of hazardous materials by fire-resistive construction, each additional space is classified as a control area. Control areas are required to be constructed in accordance with the IBC. The number of control areas allowed inside of a building and the amount of hazardous material that can be stored or used depends on the location of the storage and use in relation to the grade plane, or ground floor level of the building. In a one story building, four control areas are permitted, each containing up to the MAQ for each hazard class of the material. Footnote "a" of Table 2703.8.3.2 (See Table 16-4) allows the application of all of the quantity increases permitted in Tables 2703.1.1(1) and 2703.1.1(2), such as the installation of automatic sprinkler system throughout the building that complies with Section 903.3.1.1 and the use of hazardous material storage cabinets, gas cabinets, day boxes, and exhausted enclosures. (See Figure 16-13) [Ref. 2703.8.3.1]

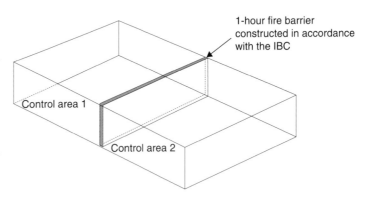

FIGURE 16-13 Control areas in a one-story building require a minimum 1-hour separation in accordance with the IBC

TABLE 16-4 Design and Number of Control Areas (IFC Table 2703.8.3.2)

Floor Level		Percentage of the Maximum Allowable Quantity Per Control Area[a]	Number of Control Areas Per Floor	Fire-Resistance Rating for Fire Barriers in Hours[b]
Above grade plane	Higher than 9	5	1	2
	7-9	5	2	2
	6	12.5	2	2
	5	12.5	2	2
	4	12.5	2	2
	3	50	2	1
	2	75	3	1
	1	100	4	1
Below grade plane	1	75	3	1
	2	50	2	1
	Lower than 2	Not Allowed	Not Allowed	Not Allowed

a. Percentages shall be of the maximum allowable quantity per control area shown in Tables 2703.1.1(1) and 2703.1.1(2), with all increases allowed in the footnotes to those tables.

b. Fire barriers shall include walls and floors as necessary to provide separation from other portions of the building.

QUESTION

A two-story Group F-1 occupancy houses a manufacturer of specialized optics. The manufacturing process requires the storage and use of a liquid hazardous material classified as a Class II organic peroxide. The building is protected throughout by an approved automatic sprinkler system in accordance with IFC Section 903.3.1.1.

The company wants to know the maximum number control areas that are allowed on each story, the MAQ of organic peroxide permitted in each control area, and the total volume of organic peroxide that can be stored inside the building.

ANSWER

Application of Table 2703.8.3.2 (See Table 16-2) allows up to seven control areas in the building—four control areas at the grade plane level and three control areas on the second floor. The MAQ on the first floor is limited to 100 pounds in each of the four control areas using the requirements in Table 2703.1.1(1), including footnote "d." The MAQ on the second floor is limited to 75 pounds in each of the three control areas. The total volume of Class II Organic Peroxide allowed in the building is 625 pounds, which must be stored so the MAQ per each control area is not exceeded.

Construction of fire barriers and horizontal assemblies that separate buildings into two or more control areas must comply with the IBC. The IBC requires the construction of fire-resistive assemblies be approved by the building official. The required fire-resistance rating of the fire

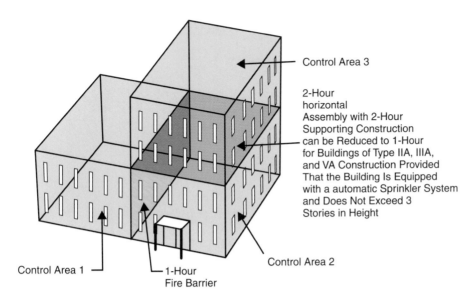

Control Area 3

2-Hour horizontal Assembly with 2-Hour Supporting Construction can be Reduced to 1-Hour for Buildings of Type IIA, IIIA, and VA Construction Provided That the Building Is Equipped with a automatic Sprinkler System and Does Not Exceed 3 Stories in Height

Control Area 2

Control Area 1

1-Hour Fire Barrier

FIGURE 16-14 The fire-resistance rating of floor areas supporting control areas in some buildings can be reduced to 1 hour when the building is protected by an automatic sprinkler system

barriers and horizontal assemblies is specified in Table 2703.8.2. As the location of the control area in relation to the grade plane increases, the fire-resistance rating of walls and horizontal assemblies also increases. When control areas are constructed four or more levels above the grade plane, the fire-resistance rating of fire barriers and horizontal assemblies increases to two hours. In buildings of Type IIA, IIIA, or VA construction three stories or less in height with control areas constructed using fire-resistive assemblies, the IBC and IFC permits the fire-resistance rating of horizontal assemblies to be reduced to one hour when the building is protected throughout by an approved automatic sprinkler system complying with the requirements in Section 903.3.1.1. (See Figure 16-14) [Ref. 2703.8.3.4]

HAZARD IDENTIFICATION SIGNS

To ensure that emergency responders are aware of the presence of hazardous materials at a facility or building, the IFC requires the posting of hazard identification signs at certain locations. A hazard identification sign must comply with the format and classification criteria in NFPA 704, *Standard System for the Identification of the Hazards of Materials for Emergency Response*. This hazard identification sign is not allowed to be used for identifying the hazards of materials in transportation. A NFPA 704 hazard identification sign identifies the health, flammability, and chemical instability of a particular hazardous material. A NFPA 704 diamond is divided into four colored fields. (See Figure 16-15) The color fields and hazards are:

- Blue – Health Hazards – 9 o'clock position
- Red – Flammable Hazards – 12 o'clock position

Code Basics

The IFC limits the number of control areas inside a building and the MAQ of hazardous materials in each control area. As the height of the control area above the grade plane level increases, the number of control areas and the MAQ on each floor is reduced. Control areas located four or more stories above the grade plane require separation using minimum 2-hour fire-resistive fire barriers and horizontal assemblies in accordance with the IBC. ●

FIGURE 16-15 The NFPA 704 identification system provides a uniform format to communicate the health, flammability, chemical instability, and special hazards to emergency responders

- Yellow – Chemical Instability – 3 o'clock position
- White – Special Hazards – 6 o'clock position

The NFPA 704 system uses a numerical scale of 0 to 4 to indicate the relative hazards of hazardous materials. A value of 0 represents the least hazardous versus a value of 4 which represent the greatest hazard. Table 16-5 summarizes the NFPA 704 hazard ratings based on the hazard categories. Additional guidance in IFC Appendix F correlates the IFC hazardous material classifications to the NFPA 704 ratings.

The IFC adopts the NFPA 704 system because it provides a uniform means of estimating the hazards of hazardous materials at a particular facility or in a given area of a building. Because NFPA 704 provides a uniform method of identifying hazards to firefighters, the system is easily implemented on local and regional levels. This is an important consideration, since hazardous materials response can involve regional deployment of emergency response personnel and equipment from multiple jurisdictions—uniformity in hazard identification is necessary to ensure hazards are accurately communicated to emergency responders. By using a standard format, many products are available in the market place that can assist businesses to easily comply with the hazard identification sign requirements. [Ref. 2703.5]

NFPA 704 hazard identification signs are required for aboveground storage tanks and stationary containers to identify the hazards of the stored material. (See Figure 16-16) Identification signs are required at specific facility entrances or locations required by the fire code official and in locations where the class and amount of hazardous material exceed the permit quantities in Section 105.6. [Ref. 2703.5]

TABLE 16-5 NFPA 704 Hazard Ratings by Hazard Categories

NFPA 704 Hazard Rating	Hazard Categories		
	Health	Flammability	Instability
4	Materials that under emergency condition, can be lethal.	Materials that rapidly or completely vaporize at atmospheric pressure and normal ambient temperature or that are readily dispersed in air and burn readily.	Materials that in themselves are readily capable of detonation or explosive decomposition or explosive reaction at normal temperatures and pressures.
3	Materials that under emergency conditions can cause serious or permanent injury.	Liquids and solids that can be ignited under almost all ambient temperature conditions.	Materials that in themselves are capable of detonation or explosive decomposition or explosive reaction but that require a strong initiating source or must be heated under confinement before initiation.
2	Materials that under emergency conditions can cause temporary incapacitation or residual injury.	Materials that must be moderately heated or exposed to relatively high ambient temperatures before ignition can occur.	Materials that readily undergo violent chemical change at elevated temperatures and pressures.
1	Materials that under emergency conditions can cause significant irritation.	Materials that must be preheated before ignition can occur.	Materials that in themselves are normally stable but that can become unstable at elevated temperatures and pressures.
0	Materials that under emergency conditions would present no hazard beyond that of ordinary combustible materials.	Materials that will not burn under typical fire conditions, including intrinsically noncombustible materials such as concrete, stone, and sand.	Materials that in themselves are normally stable, even under fire conditions.

NFPA 704 Special Hazards

A NFPA 704 special hazard designation represents hazardous materials that may be water-reactive, an oxidizer, corrosive, or a simple asphyxiation hazard. A water reactive hazard is designated with a stricken-through W (W̶). Oxidizers can be represented with the letters "OX" while corrosives are designated as "COR" or "ALK" for alkali materials. Simple asphyxiants such as inert gases or inert cryogenic fluids are designated as "SA."

FIGURE 16-16 A NFPA 704 hazard identification sign is required for aboveground storage tanks storing hazardous materials

SEPARATION OF INCOMPATIBLE MATERIALS

Accidental mixing of hazardous materials that are chemically incompatible is one cause of unauthorized discharges. It can result from an individual not understanding the hazards of mixing two chemicals together. It may result from personnel improperly modifying equipment or piping to accommodate a temporary situation where a stored energy source, such as a compressed gas, is required. Where materials have the potential to react in a manner that generates heat, fumes, gases, or byproducts that are hazardous to life or property, they are considered incompatible. Recognize that certain manufacturing processes perform intentional chemistry, which is the processing of substances so an intended chemical reaction takes place. Intentional chemistry is classified as use by the IFC, and the fire code has specific requirements to ensure that the temperature, pressure, process flow rates, and sequences of mixing are properly and safely conducted. [Ref. 2702.1 and 2705.1.11]

The provisions in Section 2703.9.8 address separating incompatible materials that may be accidentally mixed or come into contact in the event the primary container fails. The requirements are applicable to containers with a volume of more than 5 pounds or 0.5 gallons and all compressed gas containers, regardless of the cylinder or tank volume. [Ref. 2703.9.8]

A source of information for hazardous materials' chemical compatibility is the Material Safety Data Sheet (MSDS). The U.S. Department of Labor, Occupational Safety and Health Administration (OSHA) requires MSDS include information concerning the material's reactivity, which can include classes of or specific materials that are incompatible with the chemical.

Certain classes of hazardous materials are considered to be some form of a reactive hazard, which raises the potential of an unauthorized discharge. Substances with the potential of spontaneously igniting, can form or are formulated as either an inorganic or organic peroxide, water reactive materials, or self-reactive materials must be closely evaluated for incompatibility with air, water, and any contaminants that may not be readily apparent.

An electronic data base in the ICC HMEX software allows one to evaluate the incompatibility hazards if two chemicals contact or are mixed together. (See Figure 16-17) The software does not include algorithms that contemplate mixing three or more hazardous materials together. Such an operation should be treated as intentional chemistry.

The IFC offers four methods of separating incompatible materials. Separation can be accomplished by providing at least 20 feet of separation between the two incompatible materials. Solids or liquids can be located in an approved hazardous material cabinet to separate them from the other material that is incompatible. For gases, such as methane (a flammable gas) and oxygen (an oxidizer gas), one of the gas cylinders can be located into a gas cabinet or exhausted enclosure. The IFC permits incompatible materials to be located adjacent to one another, provided they are separated by a noncombustible line of sight barrier that extends a minimum of 18 inches beyond and above each stored incompatible material. (See Figure 16-18) [Ref. 2703.9.8]

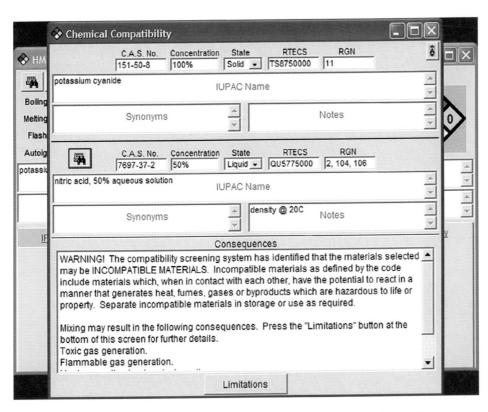

FIGURE 16-17 HMEX has a feature that allows users to evaluate if two materials mixed together are incompatible

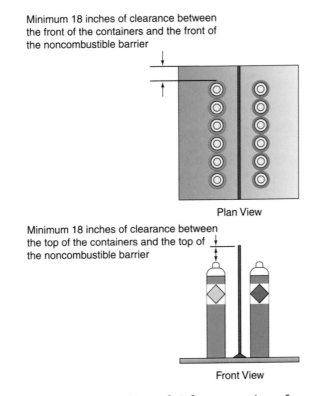

Minimum 18 inches of clearance between the front of the containers and the front of the noncombustible barrier

Plan View

Minimum 18 inches of clearance between the top of the containers and the top of the noncombustible barrier

Front View

FIGURE 16-18 Line of sight separation of incompatible hazardous materials

CHAPTER 17

Compressed Gases

(Courtesy of Air Products and Chemicals, Allentown PA)

Gases are any substance that boil at atmospheric pressure and any temperature between absolute zero (−459.7°F) and up to about 80°F. Eleven chemical elements have boiling points within this temperature range: the elements hydrogen, nitrogen, oxygen, fluorine, chlorine, helium, neon, argon, krypton, xenon, and radon. Compressed gases are divided into two major groups, depending on their physical state in containers under certain pressures and temperatures and their range of boiling points: nonliquefied gases and liquefied gases. The IFC has a third category of compressed gas known as a dissolved gas, however, this group is limited to acetylene, which is a flammable gas. Acetylene is an unstable (reactive) material and must be stabilized for safe use. Acetylene is stored in a solution of acetone to stabilize the gas. **[Ref. 3002.1]**

Nonliquefied or compressed gases do not liquefy at normal temperatures and under storage pressure which range up to about 6,000 pounds per square inch-gauge (PSIG). Compressed gases can become a liquid if cooled below their boiling points. When these gases are liquefied at these very low temperatures, they are generally referred to as cryogenic fluids and are regulated by IFC Chapter 32. Examples of compressed gases that can be converted to cryogenic fluids include nitrogen, argon, hydrogen, and oxygen.

Liquefied compressed gases become liquids at ordinary temperatures and pressures from 25 to 2,500 PSIG. Liquefied gases are elements or compounds that have boiling points relatively near atmospheric temperatures, ranging from approximately −130°F to 25–30°F. Liquefied compressed gases would become solid at the low temperatures used for storing cryogenic fluids. Liquefied gases are packaged and transported under rules that limit the maximum amount that can be packaged into a container to allow space for liquid expansion when ambient temperatures rise. Examples of liquefied compressed gases include anhydrous ammonia, propane, and carbon dioxide. (See Figure 17-1)

IFC Chapter 30 defines and sets forth requirements for the storage, use and handling of compressed gases in compressed gas containers, cylinders, tanks, and systems. The requirements in Chapter 30 are applicable to the following hazard classes of gases:

- Inert
- Corrosive
- Flammable, including liquefied petroleum gases
- Oxidizing
- Pyrophoric
- Unstable (reactive)
- Highly Toxic and Toxic
- Radioactive

Proper application and enforcement of the Chapter 30 requirements is accomplished by applying the hazardous materials specific requirements in IFC Chapters 28 through 44 and the requirements in Sections 2701 and 2703. **[Ref. 3001.1]**

The intent of the Chapter 30 requirements is to safely control the release of the potential energy stored in a compressed gas container. Gases are commonly stored at pressures of 25 to over 6,000 PSIG. The accidental release of compressed gases and its stored potential energy can be instantaneous. This rapid, violent energy release can easily propel a 125 pound gas cylinder through the air as a missile projectile—the falling cylinder and debris can injure or kill individuals.

The requirements in Chapter 30 ensure the cylinder is correctly designed and constructed to safely contain the

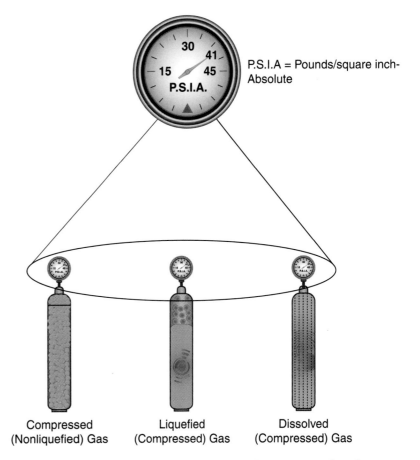

P.S.I.A = Pounds/square inch-Absolute

Compressed (Nonliquefied) Gas   Liquefied (Compressed) Gas   Dissolved (Compressed) Gas

FIGURE 17-1 Compressed, liquefied compressed and dissolved gases

FIGURE 17-2 A properly engineered and constructed compressed gas storage and distribution system

volume of filled gas at the required delivery pressure, the cylinder valve is constructed with a unique fitting to prevent it from being connected to the wrong type of pressure regulator, and the compressed gas source and system to which it is connected are operated within predetermined safe temperature, pressure and flow limits. (See Figure 17-2) The IFC requires protection of the cylinder and its connected components from mechanical impact, fire exposure, and other conditions or external exposures that could compromise the pressure containment system. The IFC prohibits the use of compressed gas containers and systems that are leaking, fire damaged, or are improperly assembled or modified.

CYLINDERS, CONTAINERS, AND TANKS

In the context of IFC Chapter 30, a compressed gas container is a pressure containment vessel. The container must be engineered to safely store and dispense the gas based on its physical or health hazards and its physical properties. Compressed gas containers are constructed and maintained in accordance with approved standards and, unless they are disposable, are subject to periodic testing and examination. Compressed gas cylinders are designed and construction in accordance with the requirements of the U.S. Department of Transportation. The IFC requires stationary unfired pressure vessels be constructed in accordance with the American Society of Mechanical Engineers *Boiler and Pressure Vessel Code* requirements. [**Ref. 3003.2**]

Regardless if it is a cylinder, container, or tank, all compressed gas containers have common and standard features:

- The size of a cylinder or container is limited. If it is a tank, like a stationary pressure vessel, its size is generally not limited.
- The volume of gas that can be stored is directly proportional to the density of the gas molecules. The lighter the gas molecules, the greater the volume of gas that can be stored when compared to heavier gas molecules stored in the same volume.
- Cylinders, containers, and tanks always have a circular cross sectional area—they are not constructed using other geometric shapes because of the imposed forces in the corners of rectangular or square shaped containers.
- Proper storage and use of compressed gas containers is dependent on its orientation. Cylinders are designed for

either a vertical or horizontal orientation— specially designed cylinders can be stored and used in either orientation.

- With the exception of a limited number of physical or health hazards materials, all compressed gas containers are equipped with a pressure relief device. The pressure relief device protects the cylinder from an explosion resulting from excessive heating of its contents.
- Valve fittings are equipped with unique, gas-specific connections to prevent the connection of the gas cylinder to the wrong pipe or distribution system.
- Cylinders require markings that identify the standard to which they are constructed, that they are qualified for service, its contents, and the basic hazards of the stored gas.

PRESSURE RELIEF DEVICES

With few exceptions, all compressed gas containers are equipped with a pressure relief device (PRD). A PRD is a safety device that protects a gas container from catastrophic failure resulting from it being overpressurized. Overpressurizing a compressed gas cylinder or tank can occur if the container is directly involved in or exposed to a fire, or if it is overfilled. PRDs provide a safe, reliable method of depressurizing gas containers so they do not explode or become a projectile. **[Ref. 3003.3.1]**

There are two forms of PRDs: reusable devices or sacrificial devices. The type of PRD selected is based on the size of the gas container and the characteristics and hazards of the stored gas. The most common reusable PRD is a spring-loaded safety relief valve. Sacrificial PRDs are either a rupture disk or a fusible plug.

Spring loaded PRDs utilize a valve that is maintained closed by a tensioned spring. The valve remains closed during normal cylinder use and filling. If the pressure inside the container exceeds a predetermined set point, the spring is compressed—this action opens a valve releasing some of the gas and lowering the internal pressure. When the container's internal pressure is reduced below the PRD set point, the valve closes, stopping the release of gas. The PRD will continue to cycle open and close until enough gas is discharged and the internal pressure is reduced. (See Figures 17–3 and 17–4)

When a sacrificial PRD operates, it cannot reclose. Operation of a sacrificial PRD expels the entire contents of the gas container. Sacrificial PRDs operate when the device's pressure or temperature limits are exceeded. Heat activated PRDs are manufactured as fusible relief valves, known as fusible plugs. Fusible

FIGURE 17-3 An internal spring-loaded safety relief valve. The valve is installed inside the gas container

FIGURE 17-4 External spring-loaded safety relief valves

plugs operate when they are subjected to direct or indirect heat from a fire. Fusible plugs are set to operate at temperatures between 130–350°F and are assembled using metals with extremely reliable and predictable melting temperatures. (See Figure 17-5)

Sacrificial pressure sensitive PRDs are known as burst discs or rupture discs. A burst disc is an engineered device that uses a metal or graphite disc that is weakened by scoring. (See Figure 17-6A) Scoring is a mechanical process where the metal is partially cut to cause a predictable failure at the opening pressure of the burst disc. The disc's size, geometry, and material of construction governs its opening or "burst" pressure. When a burst disc operates it opens like a flower and the stored gas is released to the atmosphere. (See Figure 17-6B)

Regardless of the type of PRD used, it must be properly installed and maintained. Sizing of PRDs is governed by the specification to which the container was fabricated and is based on the flow characteristics and the hazards of the stored gas. After the gas container empties, the sacrificial PRD is replaced. **[Ref. 3003.3.3]**

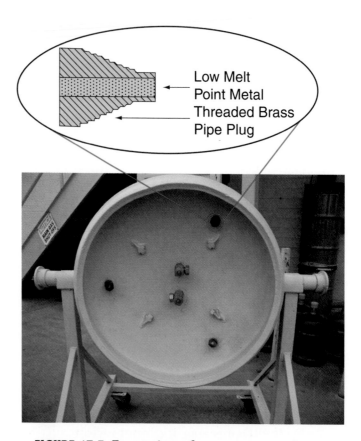

FIGURE 17-5 Front view of a one-ton capacity compressed gas container. The cylinder is equipped with three fusible plugs located at the 9 o'clock, 1 o'clock, and 5 o'clock positions. Each fusible plug is constructed by machining a hole inside a pipe plug and filling the hole with a low melt point metal

FIGURE 17-6A A burst disc installed on top of a cryogenic nitrogen container. If the container is overpressurized, the disc will burst open to safely release the cryogenic fluid

FIGURE 17-6B Unopened and opened burst discs The top burst disc has operated *(Courtesy of BS&B Safety Systems, Tulsa, OK)*

MARKINGS

Stationary and portable gas containers, cylinders, and tanks require markings. Markings demonstrate that cylinders or tanks are qualified to safely store and contain compressed gases. In the case of portable compressed gas cylinders, markings are used to identify that containers have been subjected to periodic pressure tests and inspections as prescribed by the U.S. Department of Transportation (DOT) regulations and laws. Markings communicate the hazards of the stored gas to consumers and emergency responders. When compressed gas containers are connected to a gas distribution system, markings are extended to the piping network so maintenance personnel are aware of the pipe contents and to assist fire fighters who may be called in the event of a system leak. **[Ref. 3003.4]**

Stationary containers, like pressure vessels, require markings to indentify their NFPA 704 hazard rating as well as the actual contents of the container. (See Figure 17-7) Signs are also required to indicate that open flames or other ignition sources are not permitted near these containers. **[Ref. 3003.4.1]**

Portable containers, cylinders, and tanks have various markings that identify the standard that governed their construction and design pressure. Other markings indicate that it is qualified for the compressed gas it contains. These requirements are based on the hazardous material regulations in 49 CFR Parts 100–185 promulgated by the U.S. DOT. The markings will indicate the DOT specification that the cylinder was constructed to, its type and material of construction, the cylinder's service pressure measured in PSIG, the manufacturer's mark, and the cylinder's serial number. Additional markings are required to

You Should Know

With few exceptions, all cylinders and gas containers require a means of pressure relief. Pressure relief devices are either reusable or sacrificial. Spring-loaded safety relief valves are one type of a reusable pressure relief device. Burst disks and fusible plugs are sacrificial relief devices. ●

FIGURE 17-7 Hazard identification markings on a liquid (cryogenic) nitrogen container

indicate the month and year the cylinder was manufactured and cylinder testing requirements. (See Figure 17-8)

A common consumer misconception is the color of a cylinder indicates its hazards. This assumption is not correct. A cylinder's color has no bearing on or correlation with what gas it may store. The only means of identifying the contents of a cylinder is by labeling or marking the cylinder. Cylinders require markings in accordance with Compressed Gas Association (CGA) Standard C-7, *Guide to the Preparation of Precautionary Labeling and Marking of Compressed Gas Containers*. These labels are located on the shoulder or wall of the cylinder. CGA Standard C-7 requires the marking contain the name of the hazardous material and its DOT and UN hazardous material identification label and hazard division identification number, which is used in the DOT *Hazardous Materials Emergency Response Guidebook*. (See Figure 17-9) If the compressed gas is an inhalation hazard, warning statements about

U.S. DOT Cylinder Marking information

1. Cylinder Specifications

DOT   U.S. Dept. of Transportation. (Regulatory body that governs the use of cylinders.)

3AA   Type and material of construction.

2015   Service pressure in pounds per square inch gauge.

2. A-13016   Serial number

3. SRL   Identifying symbol, registered with DOT.

4. Manufacturing Data

4-76   Date of manufacture and original test date.

H   Inspectors official mark.

+   Cylinder qualifies for 10% overfill.

*   Cylinder qualifies for 10 year retest interval.

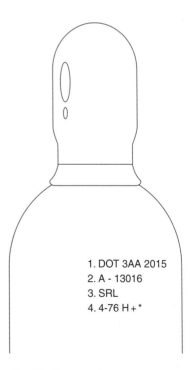

1. DOT 3AA 2015
2. A - 13016
3. SRL
4. 4-76 H + *

FIGURE 17-8 U.S. DOT required cylinder markings

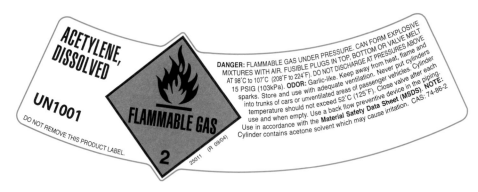

FIGURE 17-9 A CGA C-7 compliant cylinder shoulder label for a flammable gas *(Courtesy of Label Solutions, Inc.)*

this hazard are required, and its reportable quantity (RQ), if such a value is assigned by the U.S. DOT. A RQ is a value assigned by the U.S. DOT and if the amount of hazardous material released exceeds the limit, it must be reported to the United States Coast Guard National Response Center. A release of hazardous materials in amounts greater than the RQ triggers a response by the Federal government. Depending on the gas and amount of the release, the incident may result in a response by one or more federal agencies. [Ref. 3003.4.2]

Piping systems carrying compressed gases to a point of use are required to be marked with the name of the product being conveyed and the direction of flow. The markings must comply with ASME A13.1, *Scheme for the Identification of Piping Systems.* Pipe labels are required at each valve, at penetrations of a wall, floor, or ceiling, and at a maximum spacing interval of 20 feet along the entire length of the pipe run. (See Figure 17-10) [Ref. 3003.4.3]

FIGURE 17-10 Pipe labels indicate the flow and direction of anhydrous ammonia in a mechanical refrigeration system

SECURITY

To limit the likelihood of damage to compressed gas containers, the IFC requires cylinders be properly secured and protected from physical impact. The fire code prescribed three types of protection for compressed gas containers: security to safeguard the gas containers, physical protection from mechanical impact, and a method of ensuring gas cylinders are protected from being knocked over or falling. (See Figure 17-11) **[Ref. 3003.5]**

Physical protection is required when compressed gas cylinders or containers could be subjected to mechanical impact. Impact can be in the form of an object being accidentally dropped, such as the movement of materials by a crane over the compressed gas cylinders or physically striking the cylinders with material handling equipment. If containers or stationary tanks are located outdoors in an area subject to impact by vehicles, physical protection using pipe bollards or similar barriers may be warranted. (See Figure 17-12) **[Ref. 3003.5.2]**

Cylinders stored and used in either a vertical or horizontal orientation, require protection from falling as a result of being accidentally impacted, equipment vibration, or a seismic ground motion. The IFC recognizes several methods to protect cylinders from falling:

- Securing cylinders or containers to a fixed object with one or more methods of restraint or within a cylinder storage rack. (See Figure 17-13)
- Securing the cylinders or containers on a cart or other device designed for the movement of compressed gas containers, or cylinders.

FIGURE 17-11 A cylinder exchange cabinet. The cabinet is designed to protect the cylinders from tampering and is constructed of expanded metal so hose streams can apply water to cool cylinders that may be involved in a fire and provide for natural ventilation

FIGURE 17-12 Vehicle impact protection and security for a 12,000-gallon stationary pressure vessel storing a liquefied compressed gas

FIGURE 17-13 A cylinder storage rack

FIGURE 17-14 Nested cylinder storage. Note that all of the cylinders have three points of contact

- Nesting of cylinders, which is a method of securing flat-bottomed compressed gas cylinders upright in a tight mass using a contiguous three-point contact system whereby all cylinders within a group have a minimum of three points of contact with other cylinders, walls, or bracing. (See Figure 17-14) [Ref. 3003.5.3]

VALVE PROTECTION

To protect a cylinder from becoming a flying projectile, the IFC requires cylinder valves be protected when they are moved or are in storage. Valve protection is accomplished by either shielding the valve or installing a protective cap. Valve shields are installed directly on the compressed gas container and are constructed with an opening to facilitate installation or removal of pressure regulators or piping. If a cylinder is not constructed with a valve shield, it is equipped with a protective valve cap. (See Figure 17-15) [Ref. 3003.6.1]

SEPARATION FROM HAZARDOUS CONDITIONS

The IFC sets forth requirements that are intended to protect the compressed gas system from exposure to hazardous conditions. This includes incompatible hazardous materials, physical damage or excessive heating of the cylinders and containers, and ignition sources.

As previously indicated in Chapter 16, IFC Section 2703.9.8 has requirements for the separation of incompatible hazardous materials, which includes compressed and liquefied

FIGURE 17-15 A valve shield protects the valve on a 300-pound liquefied petroleum gas container

compressed gases. For certain hazard classes of gases, the fire code specifies additional requirements that further limit the potential for incompatible hazardous material storage. The IFC limits the cylinder volume of Highly Toxic and Toxic gases in Assembly, Business, Educational, Institutional, Mercantile, and Residential occupancies to 20 cubic feet and requires they be stored in a gas cabinet or exhausted enclosure. Placing these gases in gas cabinets or exhausted enclosures satisfies the requirement in Section 2703.9.8. Pyrophoric gases are separated from incompatible hazardous materials by 1-hour fire barriers unless they are stored in approved hazardous material storage cabinets, gas cabinets, or exhausted enclosures. **[Ref. 3003.7.1]**

Cylinders, containers, and tanks must be protected from excessive temperatures unless they are designed to operate in such conditions. Cylinders must be stored in an area where surface temperatures do not exceed 125°F. The intent of the requirements is to reduce the potential of the gas pressure increasing to a level that could cause the container's PRD to operate. For certain gases cylinders or their connected piping may be equipped with a device or system to heat the container. Heating the cylinder increases the pressure of the gas which allows for it to be safely distributed. Cylinder heaters are equipped with a fail safe control to ensure the temperature does not exceed 125°F and can only be used by properly trained personnel. They are installed in accordance with the requirements in the IMC and the *National Electrical Code*©. (See Figure 17-16) **[Ref. 3003.7.4, 3003.7.6]**

A common method of securing compressed gas cylinders, containers, and tanks from falling is nesting. When nesting is used on an elevated platform or near the edge of a building such as a loading dock, the gas containers must be adequately separated from the edge of the platform or shaft opening to limit the potential for damage if the cylinder or container falls. (See Figure 17-17) **[Ref. 3003.7.3]**

FIGURE 17-16 A jacket cylinder heater

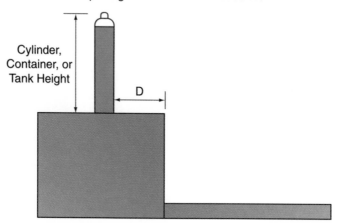

D = ½ the height of a cylinder, container or tank could fall. D is the required gas container separation distance from an unprotected ledge, platform, or elevator shaft opening. See IFC Section 3003.7.3

Cylinder, Container, or Tank Height

D

FIGURE 17-17 Cylinder near a platform

EXHAUSTED ENCLOSURES AND GAS CABINETS

Exhausted enclosures and gas cabinets are noncombustible appliances designed to capture and safely exhaust a leak from a compressed gas container or its connected piping systems. They are one of several methods permitted by Tables 2703.1.1(1) and 2703.1.1(2) that allow a 100% MAQ increase of most gases. Gas cabinets and exhausted enclosures provide a means of separating incompatible hazardous materials and comply with the requirements in Section 3003.7 for separating cylinder and containers from hazardous conditions.

The difference between an exhausted enclosure and a gas cabinet is the number of sides that are open. An exhausted enclosure is an appliance that is constructed to form a top, sides, and back and that is equipped with means of mechanical ventilation to exhaust gases or fumes. Examples of exhausted enclosure include laboratory hoods or exhausted fume hoods. (See Figure 17-18) A room or area provided with a hazardous exhaust mechanical ventilation system is not an exhausted enclosure. [Ref. 3003.7.9 and 2703.8.5]

A gas cabinet is a noncombustible enclosure that provide an isolated environment for compressed gas cylinders in storage or use. (See Figure 17-19) Gas cabinets are constructed with doors and windows that allow access to operate pressure regulators or for exchanging cylinders. They are constructed of at least 12-gage steel with self closing access ports or windows and doors and must be equipped with a mechanical ventilation system that maintains a negative pressure in relation to the surrounding room or area. Not more than three compressed gas cylinders are permitted in a single gas cabinet. [Ref. 3003.7.10 and 2703.8.6]

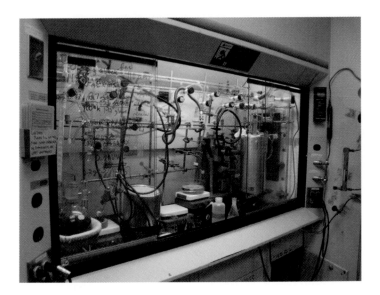

FIGURE 17-18 A laboratory hood is a type of exhausted enclosure

FIGURE 17-19 A gas cabinet

LEAKS, DAMAGE, OR CORROSION

A basic of rule of any hazardous material system, regardless of the chemical's physical state, is that leaks rarely if ever become smaller—they generally become bigger. A leak in a compressed gas system presents serious fire safety and health implications: the atmosphere in the room or area can become toxic, oxygen deficient, or can increase the possibility of a fire or explosion. Because of these risks the IFC establishes a zero-tolerance requirement that prohibits the use of leaking, damaged, or corroded gas containers. If these conditions are identified, the cylinder must be removed from service and repaired to a serviceable condition or otherwise be replaced. [Ref. 3003.12]

The term *corrosion* relates directly to the reduction of the thickness of carbon or stainless steel, such as in the wall or head of a compressed gas container, cylinder, or storage tank. Any unprotected carbon steel exposed to humidity in the air eventually will have external corrosion. Rust should not be the determining factor for whether or not the equipment should be replaced. Verification and confirmation of internal corrosion will require the measurement of the metal's thickness and a hydrostatic pressure test to determine that the container's pressure containment ability has not be reduced or compromised.

A compressed gas container that has been exposed to a fire must be removed from service. Fire exposure of a compressed gas container can cause sacrificial pressure relief devices to operate, which must be replaced. The heat of a fire can change the metallurgy of the container, which could reduce its strength or resistance to mechanical impact. Fire damaged containers, cylinders, or tanks should be removed by approved qualified personnel who have experience dealing with damaged compressed gas containers. (See Figure 17-20) [Ref. 3003.11]

FIGURE 17-20 A LP-Gas storage tank damaged by a wild land fire. This tank was removed from service, subjected to a hydrostatic pressure test, and all of its valves and fittings were replaced before it was returned to service

Flammable and Combustible Liquids

Of all the physical and health hazard materials available to private and industrial consumers, flammable and combustible liquids are the most abundant and easily accessible. Flammable and combustible liquids are used as transportation fuels, lubricants, distilled spirits, chemical feed stock in the manufacturing of plastics, pharmaceutical, semiconductors and a host of other beneficial uses.

Flammable and combustible liquids are the most regulated hazardous material in the IFC. Requirements for the storage, use, handling and dispensing are found in Chapter 34. The IFC adopts NFPA 30, *Flammable and Combustible Liquids Code* and NFPA 30A, *Motor Vehicle Fuel Dispensing Stations and Repair Garages* as additional regulations. Because flammable and combustible liquids have such a wide range of uses, other IFC chapters also have requirements when they are used for specific purposes. Other provisions that regulate specific flammable and combustible liquid uses are found in:

- IFC Chapter 6 and IMC Chapter 13 – Fuel-Fired Appliances
- IFC Chapter 11 – Aviation Facilities
- IFC Chapter 15 – Application of Flammable Finishes
- IFC Chapter 18 – Semiconductor Fabrication Facilities
- IFC Chapter 22 – Motor Fuel-Dispensing Facilities
- IFC Chapter 45 – Marinas

CLASSIFICATION OF LIQUIDS

Liquids regulated by IFC Chapter 34 are divided into two groups: Flammable Liquids and Combustible Liquids. A liquid is assigned to either group based on its boiling point and flash point temperatures. A liquid's boiling point is the temperature at which the vapor pressure of a liquid equals the absolute pressure of 14.7 pounds per square inch (PSIA). The boiling point temperature is a measurement of the volatility of the liquid and is one method of measuring how fast or slow a liquid evaporates. The lower a liquid's boiling point temperature, the more rapidly it will produce vapors. This is an important consideration because flammable and combustible liquids themselves do not burn—it is their vapors that ignite and burn. **[Ref. 2702.1]**

Flash point temperature is the minimum temperature at which a liquid will give off sufficient vapors to form an ignitable mixture with air near its surface or in the container but yet will not sustain combustion. It is the basis for classifying a liquid as either flammable or combustible. Flash point is one measure of the tendency of the liquid test specimen to form a flammable mixture with air under controlled laboratory conditions. It is only one of a number of properties that should be considered in assessing the overall flammability hazard of a liquid.

Flash point temperature is determined by tests performed in a specialized instrument. In a flash point test apparatus, a small amount of liquid is incrementally heated to a specific temperature. As the liquid temperature is raised, a pilot flame is swept directly over the test sample. If the pilot flame ignites the vapor, a flame flashes across the liquid surface. This event does not cause the vapor to continue to sustain burning and is the flash point temperature of the liquid. **[Ref. 3402.1]**

The IFC requires closed cup flash point testing be performed in accordance with one of four American Society of Testing and Materials (ASTM) standards. Closed cup tests are required by the IFC and NFPA 30 because they provide a higher degree of reproducibility. The results of closed cup flash point tests must be reproducible within ±6°F versus ±18°F for open cup flash point tests. (See Figure 18-1)

Flammable and combustible liquids are further divided into different individual classes that differentiate their flash point and boiling point temperatures. These additional categories are used to assign the relative fire hazard of liquids and serve as the basis for determining the maximum allowable quantity of liquid allowed in a control area (see Chapter 16

FIGURE 18-1 Cleveland Open Cup Flash Point test apparatus *(Courtesy of AMETEK PetroLab, Albany, NY)*

TABLE 18-1 Classification of flammable and combustible liquids

Classification	Flash point temperature	Boiling point temperature	Example liquids by class
Flammable Liquid			
Class I-A	< 73°F	< 100°F	N-Pentane; Ethyl Ether
Class I-B	< 73°F	≥ 100°F	Gasoline; Isopropyl Alcohol
Class I-C	> 73 and ≤ 100°F	Not Applicable	Methyl Ethyl Ketone
Combustible Liquid			
Class II	> 100°F and < 140°F	Not Applicable	No. 2 Diesel fuel; Kerosene
Class III-A	≥ 140°F and < 200°F	Not Applicable	Turpentine; No. 3 Fuel Oil
Class III-B	≥ 200°F	Not Applicable	Motor Oil; Cooking Oil; No.6 (bunker) Fuel Oil

for additional information). Table 18-1 presents the classification for the different classes of flammable and combustible liquids.

As previously indicated, liquids do not burn—their vapors burn. For ignition to occur, the volume of vapor must be within a given flammability range in air. The flammable range is the measurement of the percent volume of vapor that is mixed with air. Ignition can occur only when the vapor and air concentration is within the flammable range. All flammable and combustible liquids have a flammable range in air and these ranges are based on a material's lower and upper flammable limits. The lower flammable limit (LFL) is the lowest concentration at which ignition of the vapor-air mixture can occur. The upper flammable limit (UFL) is the highest concentration at which ignition of the vapor-air mixture can occur. Ignition cannot occur when the flammable vapor-air mixture is less than LFL or greater than UFL. A liquid's flammable range is based on its flash point and boiling point temperature, vapor pressure, molecular weight, and the arrangement of its molecules. Mixing of liquids will influence the flammable range as well as its flash and boiling point temperatures. The flammable range of any liquid or gas is determined by testing in accordance with one of several ASTM standards. The flammable range of liquids and gases will change if the atmosphere is enriched with oxygen or if it is diluted with an inert gas or inert liquid such as water. (See Figure 18-2)

Code Basics

Flammable and combustible liquids are classified based on their flash point and boiling point temperature. All flammable and combustible liquids exhibit a flammable range which is based on its LFL and UFL. ●

QUESTION

Carbon disulfide is used for the manufacturing of rayon and cellophane. The liquid has a boiling point temperature of 115°F and closed cup flash point temperature of 222°F. Is carbon disulfide a flammable or combustible liquid and what is its classification?

ANSWER

Class I-B flammable liquid.

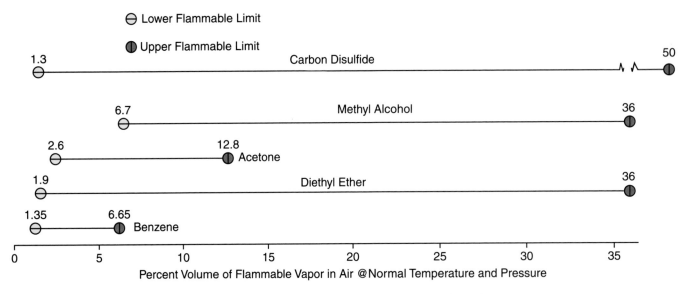

FIGURE 18-2 The flammable range of five liquids

CONTAINERS, PORTABLE TANKS, AND STATIONARY TANKS

Flammable and combustible liquids are packaged and stored in containers, portable tanks, or stationary tanks. These terms are not interchangeable, and understanding the difference between each term is an important part of code enforcement. The materials of construction, volume, and any safety features are dependent on the hazards of the liquid, specific consumer requirements such as product purity, and if the vessel is designed for transportation. If a container or portable tank can be used for transportation, it must be designed and constructed in accordance with U.S. Department of Transportation (DOT) requirements. Stationary tanks are regulated by the requirements in the IFC and NFPA 30.

The IFC limits the volume of containers and portable tanks, and NFPA 30 has specific requirements based on the classification of the liquid and the stored volume. A container is a vessel of 60 gallons or less in capacity used for transporting or storing hazardous materials. Flammable and combustible liquid containers are constructed of glass, plastic, or steel in accordance with the requirements in NFPA 30 and U.S. DOT. Pipes, piping systems, and engine fuel tanks are not considered by the IFC to be containers. NFPA 30 limits the volume of glass and plastic containers storing Class I-A, I-B, I-C, and II liquids because these containers can rapidly fail if they are subject to fire exposure. (See Figures 18-3A and 18-3B) **[Ref. 2703.2.1 and 3404.3.1]**

Portable tanks are a package of more than 60 gallons capacity that are designed for transportation. Portable tanks are equipped with components or features to facilitate material handling. Both DOT and

Figure 18-3A

Figure 18-3B

FIGURE 18-3 A and B The volume of containers is
regulated by the IFC based on the classification
of the stored liquid and its material of construction

NFPA 30 limit the volume of portable tanks to a volume of 793 gallons
(3,000 liters). A portable tank is equipped with a connection on the bottom of the tank to allow for gravity dispensing of the liquid. All portable
tanks require some means of pressure relief to prevent the tank from
rupturing or exploding if it is involved in a fire. DOT uses the term
Intermediate Bulk Container (IBC) to define many portable tanks and
permits them to be constructed of metal, plastic, fiberboard, or any combination of approved materials. DOT allows the packaging of many flammable and combustible liquids in IBCs constructed of plastic, fiberboard,
or composite materials. However, NFPA 30 prohibits the packaging of any
flammable liquids in portable tanks or IBCs constructed of these materials—
only approved metals can be used as the material of construction

FIGURE 18-4 A metallic IBC storing a flammable dye

FIGURE 18-5 Fiberboard IBC with an internal plastic bag. NFPA 30 prohibits the storage of flammable and combustible liquids in this type of package *(Courtesy of International Code Consultants, Austin TX)*

FIGURE 18-6 A stationary aboveground storage tank

Code Basics

A container has a volume of 60 gallons or less. A portable tank has a volume of 60 to 793 gallons. Containers and portable tanks are designed for transportation. A stationary tank is designed for permanent aboveground, underground, and indoor installations and generally, its size and volume is not limited. ●

when storing flammable liquids. NFPA 30 allows the packaging of combustible liquids in rigid plastic and metal IBCs—however, composite IBCs and "bags-in-a-box,"—which essentially are a plastic bag located inside of a fiberboard box—are not allowed for the storage of liquids. (See Figure 18-4 and Figure 18-5) **[Ref. 2702.1, 3404.3.1]**

A stationary tank is designed for permanent installation and is not intended for attachment to a transportation vehicle as a part of its normal operation or use. Stationary tanks are designed for aboveground or underground installation and can be installed inside of buildings. Tanks can be field-erected or fabricated in a shop and can be constructed with integral secondary containment to contain any leaks from the primary tank. Tanks are designed and installed in accordance with the IFC and NFPA 30 and with few exceptions, the size and volume of a storage tank is not limited. (See Figure 18-6) **[Ref. 2702.1, 3404.2.7]**

DESIGN AND CONSTRUCTION OF STORAGE TANKS

Flammable and combustible liquids storage tanks are designed and constructed in accordance with Underwriters Laboratories (UL) or American Petroleum Institute (API) standards as well as the requirements in

NFPA 30, NFPA 30A and the IFC. All storage tanks have features that are common among any design:

- The material of construction is chemically compatible with and designed to support the weight of the stored liquid.
- The tank has a normal vent to relieve pressures that are developed when product is introduced or withdrawn.
- Storage tanks operate at pressure of less than 15 PSIG. A stationary container designed to operate at a pressure above 15 PSIG is a pressure vessel and must be constructed in accordance with the American Society of Mechanical Engineers *Boiler and Pressure Vessel Code*.
- A nameplate is attached to the tank that indicates the standard to which it was constructed. (See Figure 18-7) [Ref. 3404.2.7]

FIGURE 18-7 Nameplate for a protected aboveground storage tank. The nameplate indicates the edition of the UL standard that tank was built to and any special installation and operation limitations

The design requirements vary based on if the tank is intended for underground or aboveground installation and its operating pressure. Tanks are designed for atmospheric pressure or low pressure service. Atmospheric tanks store liquid at atmospheric pressure (14.7 PSIA) up to 0.5 PSIG. Low pressure tanks store liquids at pressures from 0.5 to 15 PSIG. Low pressure tanks are used predominantly for storing very volatile liquids or liquids that may also be mixed with gases. [Ref. 3404.2.7]

Petroleum storage tanks fall into two categories: shop fabricated and field erected. Shop fabricated storage tanks are fabricated in a manufacturing plant and shipped as a finished assembly to the customer. Field-erected storage tanks are manufactured from plate steel cut and formed into different shapes. (See Figure 18-8) These parts are then shipped to the location where the tank is assembled and tested on site. All of the tanks listed to UL standards are shop-fabricated tanks. Tanks constructed to API standards are either shop fabricated or field-erected. However, the predominance of storage tanks constructed to API standards are field-erected.

When compared to shop-fabricated aboveground storage tanks (ASTs), field-erected ASTs are subject to more rigorous inspection requirements. The reason is that neither the IFC nor NFPA 30 limits the volume or diameter of a field-erected AST. Conversely, shop-fabricated ASTs are generally limited to around 50,000–70,000 gallons because the tank has to be assembled and tested in the factory as a condition of its listing. The finished tank is transported over highways, which limits the size and weight of loads that can be transported. Table 18-2 summarizes the major differences between shop-fabricated and field-erected ASTs.

FIGURE 18-8 Construction of a field-erected aboveground storage tank

TABLE 18-2 Design and construction differences between field-erected and shop-fabricated aboveground storage tanks

Variable	Field-erected AST	Shop-fabricated AST
Volume of Liquid	The diameter or height of the tank is not limited.	The volume is limited to between 50,000 – 70,000 gallons and transportation laws.
Design Practices	The tank bottom is the foundation and contains the thickest metal.	The tank heads are constructed of metal that is thicker than the tank shell.
Secondary Containment	A field-erected AST is normally constructed inside of an containment structure with engineered dike walls and foundation.	The tank is constructed in the factory and can be listed as having integral secondary containment.
Inspections	Field-erected ASTs are subject to periodic internal and external inspections in accordance with American Petroleum Institute standards.	The IFC does not require a periodic internal or external inspection. Liquids stored in shop-fabricated ASTs cannot be corrosive.
Approval	Is designed by a registered Professional Engineer and approved by the owner and fire code official.	Is listed by a nationally recognized testing laboratory and is approved by the fire code official.

Storage tanks are constructed to one of the following standards:

- API 620 – *Design and Construction of Large, Welded, Low-Pressure Storage Tanks*
- API 650 – *Welded Steel Tanks for Oil Storage*
- Steel Tank Institute SP01 – *Standard for Aboveground Tanks*
- UL 58 – *Steel Underground Storage Tanks for Flammable and Combustible Liquids*
- UL 142 – *Steel Aboveground Storage Tanks for Flammable and Combustible Liquids*
- UL 1316 – *Glass-Fiber-Reinforced Plastic Underground Storage Tanks for Petroleum Products*
- UL 2085 – *Protected Aboveground Storage Tanks for Flammable and Combustible Liquids*
- UL 2245 – *Vaults for Flammable and Combustible Liquid Tanks*

Underground storage tanks (USTs) are constructed of carbon steel or fiberglass-reinforced thermosetting plastic and are designed for burial. USTs provide the highest level of fire safety because the tank has no possibility of being exposed to a fire. (See Figure 18-9) Because a UST is buried, many states require these tanks to be constructed with a primary and secondary containment so in the event of a leak, the outer (secondary) containment shell will prevent the petroleum product from contaminating soil or a water body. These tanks are equipped with an overfill protection system and electronic systems

FIGURE 18-9 Installation of an underground storage tank (*Courtesy of The Steel Tank Institute, Lake Zurich, IL*)

for leak detection and product reconciliation. Product reconciliation is an important part of an overall spill prevention program as it ensures each gallon of petroleum product is accounted for through receipt, sales, or evaporative losses. If the UST is constructed of steel, it is connected to a cathodic protection system to protect it from corrosion.

Aboveground storage tanks (ASTs) are constructed of carbon steel or other ferrous materials. (See Figure 18-10) They can be installed outside or inside of buildings. Because an AST can be easily inspected, it normally does not require cathodic protection. Based on the type of tank and its application, tank connections, known as nozzles, may be above or below the liquid level. Even though ASTs can be visually inspected, almost all installations will require secondary containment. This protection is required to prevent any leaks from forming a large spill of petroleum product, which could be the fuel for a pool fire. Except for tanks larger than 12,000 gallons storing Class III-B combustible liquids, every AST requires an emergency vent, which protects the tank from a pressure explosion if it is involved in or an exposure to fire.

FIGURE 18-10 An aboveground storage tank at a manufacturing plant

ASTs and USTs are commonly used for the storage of vehicle fuels including unleaded gasoline, gasoline-ethanol mixtures, diesel fuel, and biodiesel. ASTs are commonly specified by many petroleum product suppliers and retailers because they are subject to fewer permit fees that fund government-mandated leaking UST payment programs. The requirements for the type of tank allowed for motor fuel-dispensing facilities are specified in IFC Chapter 22. Since many motor fuel-dispensing facilities are accessible by the general public, the IFC prescribes the use of either ASTs located in below grade vaults or Protected ASTs when storing Class I liquids above ground and outside of buildings. When a Protected AST is used, the tank must be installed in accordance with the requirements in Chapter 34. [Ref. 2206.2.3]

A protected AST (PAST) is a shop fabricated tank listed in accordance with UL 2085,

FIGURE 18-11 A 10,000 gallon protected aboveground storage tank at a fleet vehicle motor fuel-dispensing facility

Protected Aboveground Storage Tanks for Flammable and Combustible Liquids. A PAST consists of a primary tank that is constructed to the requirements of UL 142 and is protected from physical damage such as impact by a vehicle. (See Figure 18-11) PASTs are designed to operate at atmospheric pressure and are constructed with a method of secondary containment that it is integral to the storage tank. The tank is constructed with a fire-resistive feature to protect it from a liquid petroleum pool fire. The tank may provide protection elements as a unit, may be an assembly of components, or may be a combination of both. [Ref. 3404.2.9.7]

TABLE 18-3 Minimum separation requirements for aboveground tanks (IFC Table 2206.2.3)

Class of liquid and tank type	Individual tank capacity (gallons)	Minimum distance from nearest important building on same property (feet)	Minimum distance from nearest fuel dispenser (feet)	Minimum distance from lot line that is or can be built upon, including the opposite side of a public way (feet)	Minimum distance from nearest side of any public way (feet)	Minimum distance between tanks (feet)
Class I protected aboveground tanks	Less than or equal to 6,000	5	25[a]	15	5	3
	Greater than 6,000	15	25[a]	25	15	3
Class II and III protected aboveground tanks	Same as Class I	Same as Class I	Same as Class I	Same as Class I	Same as Class I	Same as Class I
Tanks in vaults	0 – 20,000	0[b]	0	0[b]	0	Separate compartment required for each tank
Other tanks	All	50	50	100	50	3

For SI: 1 foot = 304.8 mm, 1 gallon = 3.785L

a. At fleet vehicle motor fuel-dispensing facilities, no minimum separation distance is required.

b. Underground tanks shall be located such that they will not be subject to loading from nearby adjacent structures, or they shall be designed to accommodate applied loads from existing or future structures that can be built nearby.

Installation requirements for ASTs are found in IFC Chapters 22 and 34. IFC Table 2206.2.3 (See Table 18-3) sets forth the requirements for locating the tank based on the type of tank, its volume, and its location in relation to certain exposures, including important buildings on the same property, the dispenser, property lines, and public ways. [Ref. 2206.2.3]

When a PAST is used at a motor fuel-dispensing facility it also must comply with the requirements in NFPA 30A and IFC Chapter 34. The PAST also must be equipped with an overfill prevention device and a spill container. (See Figure 18-12) An overfill prevention device prevents the PAST from being overfilled with flammable or combustible liquids. The device is connected to a liquid-tight tank fill connection. An overfill prevention device is designed with a valve that floats on the liquid inside the tank. As the liquid level is increased during filling, the valve begins to restrict the liquid when the PAST is filled to 90% of its rated volume. The valve continues to restrict the flow until the PAST is filled to 95% of its rated volume—at this point, the flow of liquid is stopped by the valve. To further limit the potential for a flammable or combustible

FIGURE 18-12 A spill container installed on a protected aboveground storage tank. It encloses the tank fill connection, which is equipped with an overfill prevention device

liquid release, the fill connection is located inside of a spill container. The spill container is designed to capture any liquid that remains in the hose supplying the fuel from a cargo tanker. A spill container is designed to retain at least 5 gallons of liquid and is equipped with a valve to drain spilled liquid into the PAST. [**Ref. 3404.2.9.6.6, 3404.2.9.6.8**]

STORAGE TANK OPENINGS

A tank has several openings to accommodate simultaneous tank filling and product withdrawal as well as other components. Components installed on storage tanks must comply with the applicable IFC and NFPA requirements. USTs and ASTs are equipped with at least two openings. One opening is designed to allow the tank to vent vapor to the atmosphere when product is introduced into the storage tank. It allows outside air inside of the tank when product is withdrawn during transfer or dispensing. This opening is called the normal vent. A second opening is provided for introducing or withdrawing product from the tank. [**Ref. 3404.2.7.3 and 3404.2.7.5**]

Storage tanks are designed to resist the vacuum and positive pressures generated when liquid is introduced into or withdrawn. Improperly sizing a tank's normal vent or obstruction it can generate excessive negative pressure, causing the tank to collapse into itself. Normal vents on storage tanks storing Class I, II, and III-A liquids are terminated outside of buildings at least 12 feet above the adjacent ground level. The 12-foot elevation of the vent is necessary to ensure that the surrounding air mixes with the vapor being exhausted to ensure it is diluted below 25%

Code Basics

Stationary storage tanks are constructed in accordance with recognized national standards approved by the fire code official. Underground storage tanks are shop-fabricated. Aboveground storage tanks are field-erected or shop-fabricated. Shop-fabricated tanks are listed by a nationally recognized testing laboratory, while field-erected tanks are approved by the fire code official and the owner of the storage tank. •

of the liquid's LFL. Vent piping are arranged so vapors are discharged upward or away from property lines. The normal vent is terminated at least 5 feet from building openings or lot (property) lines that can be built upon. (See Figures 18-13A and 18-13B) **[Ref. 3404.2.7.3.3]**

The normal vent on a tank storing Class I liquids must be equipped with a pressure-vacuum (PV) vent. (See Figure 18-14) A PV vent is designed to remain closed during normal operation—the vent only operates when the tank is operating under a positive pressure or negative pressure (vacuum) conditions. The IFC specifies a PV vent to minimize the loss of vapor when the tank is atmospherically heated. **[Ref. 3404.2.7.3.6]**

A third opening is provided for all ASTs storing Class I, II, and III-A liquids and tanks storing less than 12,000 gallons of Class III-B liquids. The emergency vent opening will safely vent vapors that will be generated if the tank is involved in a hydrocarbon pool fire or is subject to an exposure fire. An emergency relief vent is an opening, a construction method, or device that relieves internal pressure that can develop inside of a storage tank when it is exposed to fire. An emergency vent is not required for ASTs that contain more than 12,000 gallons of Class III-B liquids provided the tank is not located with the same containment dike or a drainage path of Class I or II flammable or combustible liquids. **[Ref. 3404.2.7.4]**

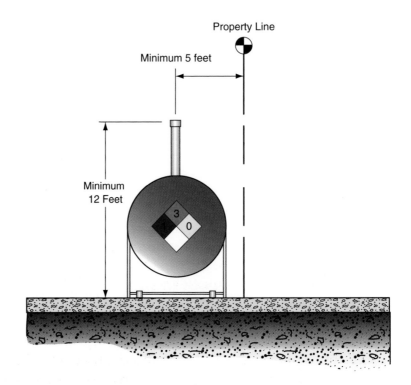

FIGURE 18-13A Normal venting of an aboveground storage tank

A Storage Tank is Designed to

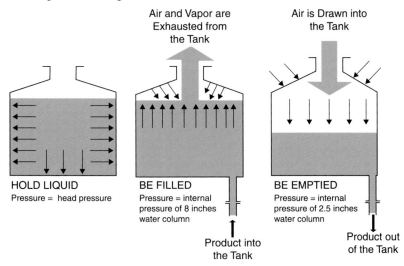

Air and Vapor are
Exhausted from
the Tank

Air is Drawn into
the Tank

HOLD LIQUID
Pressure = head pressure

BE FILLED
Pressure = internal
pressure of 8 inches
water column

Product into
the Tank

BE EMPTIED
Pressure = internal
pressure of 2.5 inches
water column

Product out
of the Tank

What Are Inches Water Column?

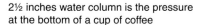

2½ inches water column is the pressure
at the bottom of a cup of coffee

8 inches water column is the pressure at
the bottom of a glass of beer

FIGURE 18-13B Functions of the normal vent

A number of notable incidents have occurred in the United States involving the failure to provide or disabling of emergency vents on ASTs that caused the death of firefighters. One of the more devastating incidents was a fire at a small petroleum terminal and motor fuel-dispensing station that occurred August 18, 1959 in Kansas City, Kansas. The fire involved a nominal 20,000-gallon horizontal AST. The tank was not listed nor found to be constructed in accordance with any API or UL standards. The tank had a 1½-inch normal vent but was not equipped with an emergency vent. Approximately 90 minutes into the firefighting operations, the storage tank catastrophically failed because of the internal pressure and the weakening of one end of the storage tank that was not being cooled by fire streams. When the tank failed, it propelled the storage tank through two brick walls and approximately 95 feet into the firefighters' operating large hose streams onto the fire. The fire and tank failure killed five firefighters, one civilian, and injured over 65 persons. This was largest number of line of duty deaths experienced at that time by

FIGURE 18-14 Pressure-vacuum vents on each of the three normal vent pipes connected to underground storage tanks

the Kansas City (KS) Fire Department. Calculations of the wetted surface area of the tank found that it should have been equipped with a 10-inch emergency vent.

On July 31, 1968, a similar incident occurred outside of Kennadale, Texas. The fire involved a 10,000-gallon home made AST storing gasoline. The tank was divided into two compartments—7,000 and 3,000 gallons respectively. The 7,000 gallon compartment was equipped with an over-fill opening that acted as an emergency relief vent—however the size of the opening was undersized by 50%. The 3,000 gallon compartment did not have an emergency vent. Approximately one hour into the incident, the storage tank exploded, killing two firefighters, one bystander, and injuring 57 firefighters and civilians.

Several methods are available for satisfying the IFC requirements for emergency venting of ASTs. On shop-fabricated ASTs, the most common methods are the installation of a direct action emergency vent or a long-bolt emergency vent. A direct action emergency vent is constructed to a minimum diameter and a cover that is attached to a metal guide. (See Figure 18-15) The internal diameter of the direct action emergency vent governs its pressure relief flow capacity. The weight of the cover governs its opening pressure, which cannot exceed 2.5 PSIG under the NFPA 30 requirements. Direct-action emergency vents offer a reliable means of emergency venting for ASTs because the cover is the only moving part.

Another acceptable method of providing an emergency vent is installing a long-bolt vent. (See Figure 18-16) A long-bolt emergency vent is simply a flange cover mounted onto a flanged man way opening on top of an AST. Every other bolt is removed from the flange cover and the remaining bolts are installed so they have minimum 1½ inches of

FIGURE 18-15 Direct action emergency vent. This particular manufacturer uses a spring-loaded cover that is secured closed by a graphite pin. The pin breaks at an opening pressure of approximately 1 PSIG, causing the vent cover to open

FIGURE 18-16 Long-bolt emergency vent

movement. These vents are less reliable because the cover can be bolted to the flange man way opening which disables it. They are also subject to warping, especially in hot climates when the metal is cooled by rain.

To determine the adequacy of an emergency vent, NFPA 30 requires they be marked to indicate the vent's flow rate. The flow rate, expressed in standard cubic feet/hour (SCFH) is based on the wetted area of the AST. The wetted area is the internal tank area made wet by the stored liquid. Calculation of the wetted area is based on the shape of the storage tank and if it is designed for a horizontal or vertical installation.

The nameplate for UL listed ASTs are marked to indicate the emergency vent flow rate. An inspection of any AST should include a comparison of the emergency vent flow rate to the required flow rate on the tank nameplate. An emergency vent is undersized if the flow rate does not equal or exceed the rated flow capacity on the storage tank nameplate. NFPA 30 permits adding the rated flow capacity through the normal vent to the rated capacity of the emergency vent to satisfy the total required flow rate indicated on the tank nameplate. However, the normal vent must be marked to indicate its rated flow capacity. (See Figure 18-17)

FIGURE 18-17 The flow rate of an emergency vent must be marked on the device

GLOSSARY

A

addition – An extension or increase in floor area, number of stories, or height of a building or structure.

alteration – Any construction or renovation to an existing structure other than a repair or addition.

appliance – Any apparatus or equipment that utilizes gas as a fuel or raw material to produce light, heat, power, refrigeration or air conditioning.

approved – Acceptable to the fire code official.

automatic sprinkler system – For fire protection purposes, an integrated system of underground and overhead piping designed in accordance with fire protection engineering standards. The system includes a suitable water supply. The portion of the system above the ground is a network of specially sized or hydraulically designed piping installed in a structure or area, generally overhead, and to which automatic sprinklers are connected in a systematic pattern. The system is usually activated by heat from a fire and discharges water over the fire area.

B

boiling point – The temperature at which the vapor pressure of liquid equals the atmospheric pressure of 14.7 pounds per square inch (PSIA) or 760 mm of mercury. Where an accurate boiling point is unavailable for the material in question, or for mixtures which do not have a constant boiling point, for the purposes of this classification, the 20-percent evaporated point of a distillation performed in accordance with ASTM D86 shall be used as the boiling point of the liquid.

branch circuit – That part of an electric circuit extending beyond the last circuit breaker or fuse. The branch circuits start at the breaker box and extend to the electrical devices connected to the service. Branch circuits are the last part of the circuit supplying electrical devices. These circuits are classified in two different ways, according to the type of loads they serve or according to their current-carrying capacity.

C

catalyst – A substance that initiates or accelerates a chemical reaction without itself being affected.

change of occupancy – In the purpose or level of activity within a building that involves a change in application of the requirements of the *International Existing Building Code*.

closed use system – The use of a solid or liquid hazardous material involving a closed vessel or system that remains closed during normal operations where vapors emitted by the product are not liberated outside of the vessel or system and the product is not exposed to the atmosphere during normal operations; and all uses of compressed gases.

code – A written set of rules, or principles, or laws.

combustible dust – Finely divided solid material which is 420 microns or less in diameter and which, when dispersed in air in the proper proportions, could be ignited by a flame, spark or other source of ignition. Combustible dust will pass through a U.S. No. 40 standard sieve.

commodity – A combination of products, packing materials and containers.

construction documents – The written, graphic and pictorial documents prepared or assembled for describing the design, location and physical characteristics of the elements of the project necessary for obtaining a permit.

corrosive – A chemical that causes visible destruction of, or irreversible alterations in, living tissue by chemical action at the point of contact. A chemical shall be considered corrosive if, when tested on the skin of albino rabbits by the method described in DOTn 49 CFR 173.137, such chemical destroys or changes irreversibly the structure of the tissue at the point of contact following an exposure period of 4 hours. The term does not refer to action on inanimate surfaces.

D

deflagration – An exothermic reaction, such as the extremely rapid oxidation of a flammable dust or vapor in air, in which the reaction progresses through the unburned material at a rate less than the velocity of sound. A deflagration can have an explosive effect.

dispensing – The pouring or transferring of any material from a container, tank or similar vessel, whereby vapors, dusts, fumes, mists or gases are liberated to the atmosphere.

E

emergency voice/alarm communications – Dedicated manual or automatic facilities for originating and distributing voice instructions, as well as alert and evacuation signals pertaining to a fire emergency, to the occupants of a building.

exit – That portion of a means of egress system which is separated from other interior spaces of a building or structure by fire-resistance-rated construction and opening protectives as required to provide a protected path of egress travel between the exit access and the exit discharge. Exits include exterior exit doors at the level of exit discharge, vertical exit enclosures, exit passageways, exterior exit stairways, exterior exit ramps and horizontal exits.

exit access – That portion of a means of egress system that leads from any occupied portion of a building or structure to an exit.

exit discharge – The portion of a means of egress system between the termination of an exit and a public way.

F

fire – A chemical reaction that releases heat and light and is accompanied by flame, especially the exothermic oxidation of a combustible substance.

fire area – The aggregate floor area enclosed and bounded by fire walls, fire barriers, exterior walls or horizontal assemblies of a building. Areas of the building not provided with surrounding walls shall be included in the fire area if such areas are included within the horizontal projection of the roof or floor next above.

fire code official – The fire chief or other designated authority charged with the administration and enforcement of the code, or a duly authorized representative.

fire protection system – Approved devices, equipment and systems or combinations of systems used to detect a fire, activate an alarm, extinguish or control a fire, control or manage smoke and products of a fire or any combination thereof.

fire resistance – That property of materials or their assemblies that prevents or retards the passage of excessive heat, hot gases or flames under conditions of use.

fire flow – The flow rate of a water supply, measured at 20 pounds per square inch (psi) residual pressure that is available for fire fighting.

flashover – Is an event during a fire's growth where the hot smoke layer inside a room or compartment releases the greatest amount of convective and radiant energy.

H

heat release rate – A measurement of the rate a combustion reaction produces heat and calculated by multiplying the effective heat of combustion of a material by its mass loss rate. It is expressed in British Thermal Units [BTU]/minute or kilowatt (KW).

high-piled combustible storage – Materials in closely packed piles or combustible materials on pallets, in racks or on shelves where the top of storage is greater than 12 feet in height. When required by the fire code official, high-piled combustible storage also includes certain high-hazard commodities, such as rubber tires, Group A plastics, flammable liquids, idle pallets and similar commodities, where the top of storage is greater than 6 feet in height.

hood – An air intake device used to capture by entrapment, impingement, adhesion, or similar means, grease, moisture, heat and similar contaminants before they enter a duct system.

I

inspection – A formal or official examination.

interlock – A method of preventing undesired states in a machine, which in a general sense can include any electrical, electronic, or mechanical device or system.

L

level of exit discharge – The story at the point at which an exit terminates and an exit discharge begins.

liability – The state of being legally obliged and responsible.

iower flammable limit (LFL) – The minimum concentration of vapor in air at which propagation of flame will occur in the presence of an ignition source. The LFL is sometimes referred to as LEL or lower explosive limit.

M

manometer – An instrument for measuring differences of pressure. The weight of a column of liquid enclosed in a tube is balanced by the pressures applied at its opposite ends, and the pressure difference is computed from the hydrostatic equation.

O

occupant load – The number of persons for which the means of egress of a building or a portion thereof is designed.

open use system – The use of a solid or liquid hazardous material involving a vessel or system that is continuously open to the atmosphere during normal operations and where vapors are liberated, or the product is exposed to the atmosphere during normal operations.

P

performance-based design – An engineering approach to design elements of a building based on agreed upon performance goals and objectives, engineering analysis and quantitative assessment of alternatives against the design goals and objectives using accepted engineering tools, methodologies and performance criteria.

pressure, residual – The available pressure inside of pipe while the liquid is flowing.

pressure, static – The pressure inside of a pipe or container while the liquid is at rest.

R

retroactive – Affecting things past.

S

spray booth – A mechanically ventilated appliance of varying dimensions and construction provided to enclose or accommodate a spraying operation and to confine and limit the escape of spray vapor and residue and to exhaust it safely.

spraying space – An area in which dangerous quantities of flammable vapors or combustible residues, dusts or deposits are present due to the operation of spraying processes. The fire code official is authorized to define the limits of the spraying space in any specific case.

standard – Something set up and established by authority as a rule for the measure of quantity, weight, extent, value or quality.

storage – The keeping, retention, or leaving of hazardous materials in closed containers, tanks, cylinders, or similar vessels; or vessels supplying operations through closed connections to the vessel.

supervisory signal – A signal indicating the need of action in connection with the supervision guard tours, the fire suppression system or equipment, or the maintenance features of related systems.

W

wildland – An area in which development is essentially nonexistent, except for roads, railroads, power lines and similar facilities.

INDEX

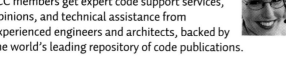

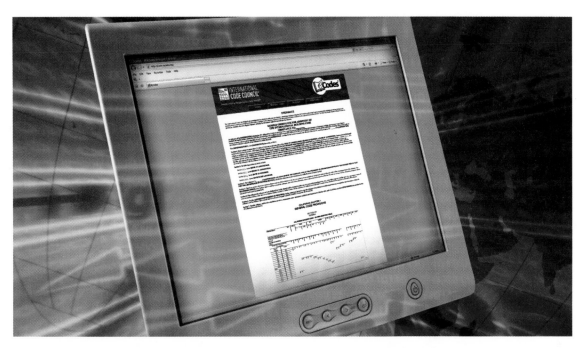